TACTICS AND TECHNIQUES OF ELECTRONIC WARFARE

Electronic Countermeasures in the Air War Against North Vietnam 1965-1973

by

Bernard C. Nalty

DEFENSE LION PUBLICATIONS

FOREWORD

Electronic Countermeasures in the Air War Against North Vietnam is one of a series of monographs on USAF tactics and techniques in Southeast Asia. Electronic countermeasures is but one aspect of the broad subject of electronic warfare, which was waged in all its complexity throughout Southeast Asia. Nevertheless, in choosing to deal with this topic, the Office of Air Force History faced a difficult security problem, for key material is so closely held that its inclusion might result in a history to which the average Air Force officer would not have access. As a result, this special intelligence was not used. Instead, various agencies involved in electronic countermeasures, having access to material not available to the historian, were invited to comment on a draft of the monograph, to ensure an essentially correct account.

The author did most of his research at the Office of Air Force History, using materials obtained from the Albert F. Simpson Historical Research Center and the Air University Library, Maxwell AFB, Ala. The U.S. Air Force Security Service, (USAFSS, San Antonio, Tex., furnished certain countermeasures evaluations that did not contain unusually sensitive information. Data from this organization supplemented the material assembled by the Strategic Air Command, Offutt AFB, Neb., on the 1972 B-52 campaign against North Vietnam.

JOHN W. HUSTON, Maj. Gen. USAF

Chief, Office of Air Force History

Washington, D.C.

CONTENTS

PREFACE

Unlike the broader subject of electronic warfare, which originated with interceptions of radio traffic during World War I, electronic countermeasures began with the appearance of radar-directed air defenses in World War II. The first systematic use of electronic countermeasures occurred when British night bombers employed various devices to blind German radar and disrupt communications between defending pilots and ground controllers. U.S. Army Air Forces also conducted wartime countermeasures operations, and during the Korean fighting the U. S. Air Force used equipment and techniques developed for World War II. In the years that followed, the United States sought to keep pace with improvements in radar by devising new countermeasures, especially for the strategic bomber force, though for tactical aircraft as well.

The Vietnam war tested the recent developments in electronic countermeasures. At first, radar-controlled surface-to-air missiles and anti-aircraft guns had the advantage. The Air Force. however, perfected a countermeasures pod for fighter-bombers, and fitted out and armed aircraft for the express purpose of locating and destroying missile sites. These endeavors, complemented by long-range jamming and by countermeasures aircraft from the other services, succeeded in restoring a balance favorable to the offense.

The deadly struggle continued throughout the war. The North Vietnamese adjusted their radars and electronic techniques to neutralize American countermeasures, and the Americans reacted to the changing threat. The countermeasures effort reached its climax in Linebacker II, the B-52 attacks of December 1972 against the Hanoi-Haiphong region. The entire wartime experience was compressed into a few days, as each side sought to overcome the electronic tactics employed by the other.

The electronic countermeasures campaign produce no inflexible catechism of lessons learned. Instead, the Air Force discovered the importance of continually evaluating the usefulness of its countermeasures and adjusting quickly when effectiveness declined. The basic lesson can best be summed up in words attributed to Adm. Thomas Moorer, former Chairman of the Joint Chiefs of Staff: "If there is a World War III, the winner will be the side that can best control and manage the electromagnetic spectrum."

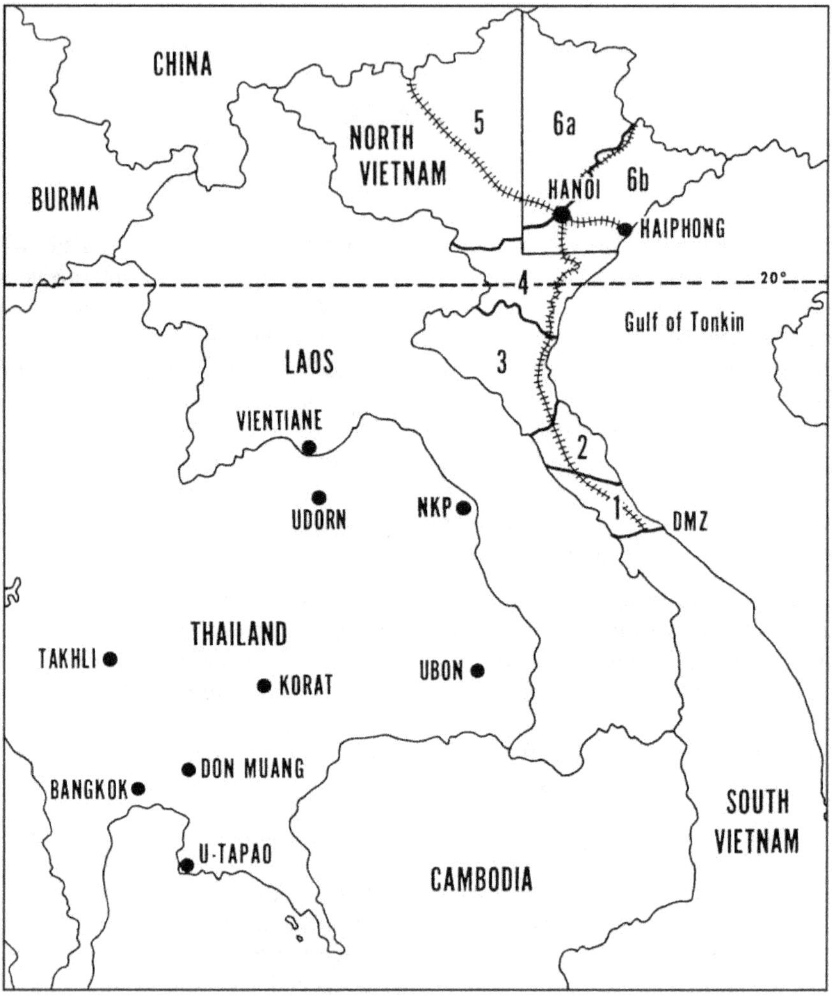

Mainland
Southeast Asia

I. THE DESTRUCTION OF LEOPARD 02

The four McDonnell Douglas F-4C Phantoms of Leopard flight were cruising in loose fingertip formation at 23,000 feet, some 37 nautical miles west of Hanoi, capital city of North Vietnam. These twin-jet fighter-bombers, each with a 2-man crew, had taken off earlier that morning from Ubon air base, Thailand, to protect other Air Force planes assigned to bomb North Vietnamese military targets. The date was 24 July 1965, during the fifth month of Rolling Thunder, an air offensive designed, according to President Lyndon B. Johnson, to make North Vietnam's leaders "realize... that their aggression" against neighboring South Vietnam "should cease." [1]

The RB-66C's warning came too late . . . Source: U.S. Air Force

At about 0805 local time, a Douglas RB-66C electronic reconnaissance plane detected a radar signal from a Russian-designed SA-2 surface-to-air missile battery, the second time in as many days that one of these planes had intercepted this type of transmission originating within North Vietnam. The RB-66C radioed a warning, and shortly afterward Lt. Col. William A. Alden, USAF on the far left of the Leopard formation, saw two missiles streaking skyward, closing rapidly from the right and below. The first missile exploded directly beneath Leopard 02, on the opposite flank from Alden. He saw flames erupt from the trailing edge of the Phantom's wing, as the stricken plane rolled onto its back and spiraled out of sight into the clouds. One of the crew, Capt. Richard P. Keirn, parachuted, survived almost eight years as a prisoner of war, and returned to the United States in 1973. His partner, Capt. Roscoe H. Fobair, apparently died in the crash. Alden and the surviving members of Leopard flight broke formation, evaded the second missile, and returned to Ubon, where ground crews discovered that all three aircraft bore scars from fragments of the missile that had downed Leopard 02. [2]

A North Vietnamese missile battery crew at work. Source: People's Liberation Army of Vietnam

Introducing the SAM [3]

Well before the Vietnam conflict the American military coined an acronym for the surface-to-air missile; it had become the SAM. Although the SAM made a grim Southeast Asia debut, destroying a $2-million airplane and damaging three others, it was not a new weapon. The SAM had first appeared in 1957, and 3 years later it shot down an American Lockheed U-2 high-altitude reconnaissance plane near Sverdlovsk, deep inside the Soviet Union. The pilot, Francis Gary Powers of the Central Intelligence Agency, was taken prisoner, but within 2 years his captors released him in ' 1 exchange for a Russian spy held in the United States. Less fortunate was the Chinese Nationalist pilot killed in the autumn of 1962 when a SAM site in mainland China downed his U-2. A few weeks later, during the Cuban missile crisis, this type of weapon destroyed a Strategic Air Command (SAC) U-2 on a photographic mission over Cuba, killing the pilot, Maj. Rudolf Anderson, Jr. [4]

The kind of SAM that destroyed Leopard 02 was a Guideline missile, a component of the SA-2 weapon system. The SA-2 employed a Fan Song radar to locate targets for four to six Guideline missiles mounted on individual launchers. Portable generators provided current for the computers that processed firing data. At each SAM site, the North Vietnamese also installed an acquisition radar capable of detecting aircraft at a distance of 100 nautical miles, roughly three times the range of Fan Song. The most common kind of acquisition radar was called Spoon Rest, though others saw service before the war ended.

The Guideline missiles that roared skyward on 24 July measured 10.6 meters (35 feet) in length, weighed about 2270 kilograms (5000 pounds), and attained a velocity approaching Mach 4. The solid-propellant booster rocket was 64 centimeters (26 inches) in diameter and 2.6 meters (8.5 feet) in length. The liquid-fueled sustainer, or second stage, measured 49 centimeters (20 inches) in diameter and was 7.9 meters (26 feet) long. Because of its distinctive shape, Guideline received the nickname "flying telephone pole" The high-explosive warhead weighed 189.6 kilograms (420 pounds) and had a lethal bursting radius of 150 to 200 feet. Although Guideline could down an airplane as far as 17 nautical miles from the launch site, it could not engage targets within a "dead zone" of 5 nautical miles. The missile had to travel this distance before the fuse was armed, the booster stage discarded, and the guidance system in operation. Nor was Guideline effective against aircraft flying below 3000 feet, the minimum altitude at which Fan Song could track a fast-flying target.

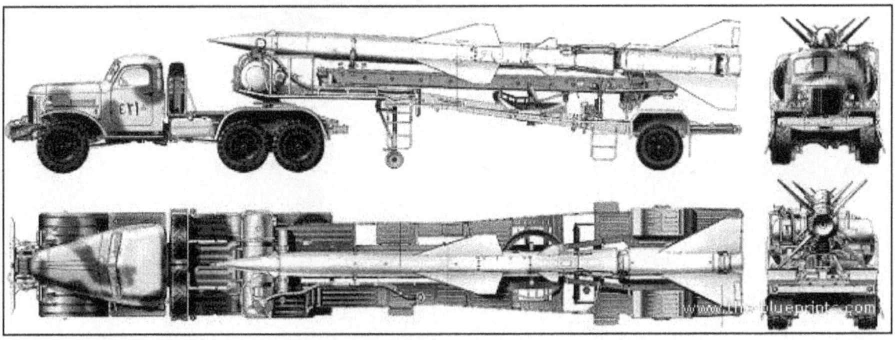

SA-2 Missile on a towed trailer. The Soviet designation for this weapon was the S-75 Dvina. Source: U.S. Air Force.

Mounted on a trailer, Fan Song employed two trough-shaped antennas, one of which scanned horizontally to determine azimuth and range, while the other scanned vertically to obtain elevation and verify range. Because the antennas continued to scan after they had locked onto a target, Fan Song was called a track-while-scan radar. This feature enabled the operator to detect additional targets while tracking a particular radar return but did not permit him to fire upon two or more targets simultaneously. To acquire a target, lock onto it, and launch a missile took about 75 seconds, with 30 to 40 seconds required to shift to another target, lock on, and launch. SAM crews could salvo their missiles against a single radar return at 5-second intervals.

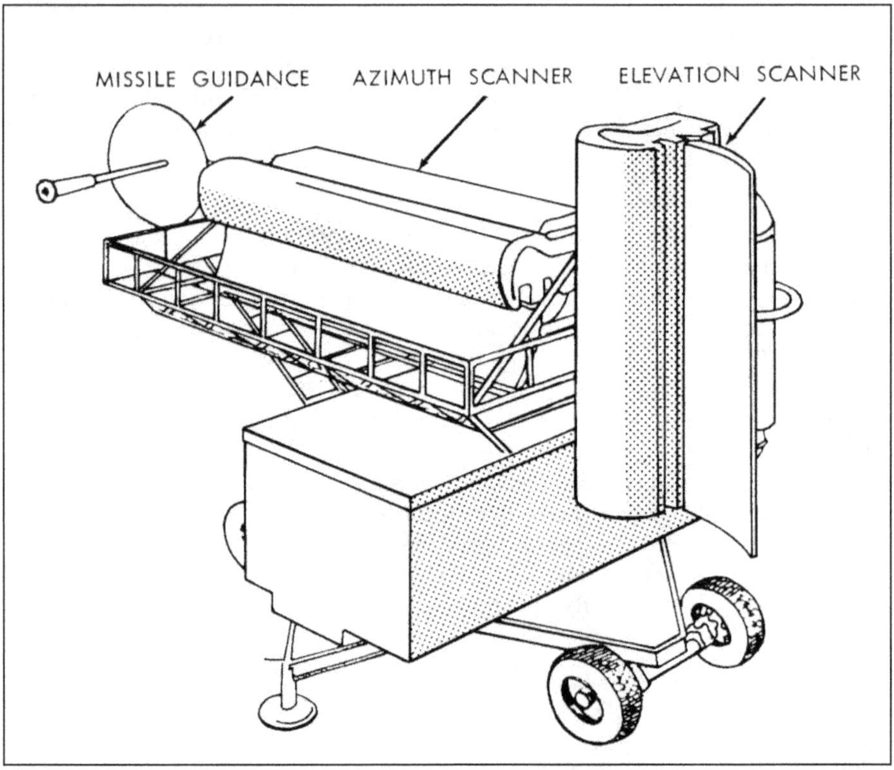

The Fan Song Radar. This had the Soviet designation SNR-75. Source: U.S. Air Force.

The same antennas that tracked the target also tracked the missile, receiving a signal from the guidance beacon, a transponder located on the sustainer stage. A computer processed the data on both missile trajectory and flight path of the target, issuing commands to the missile by means of a guidance transmitter that used a different frequency from the track-while-scan radar and had its own dish-shaped antenna. The guidance signal which commenced no later than 4 seconds after launch, was picked up by a receiver in the base of Guideline's sustainer section.

When the SAM was about 1000 feet from the calculated point of interception, or approximately 23 seconds after launch, a command from the ground armed the warhead. A radar proximity fuse caused detonation. If the intended victim eluded the missile, the warhead would automatically explode from 57 to 63 seconds after launch. The second Guideline fired on 24 July against Leopard flight destroyed itself in this manner.

The SA-2 was a mobile weapon. Some 25 assigned vehicles and vans could move a complete battalion, usually consisting of an acquisition radar, a Fan Song set, electrical generators, a fire control computer, and 18 Guideline missiles, one for each of the six launchers and a dozen spares. At first glance, rapid movement seemed almost impossible in the absence of an extensive network of surfaced roads, but the North Vietnamese proved surprisingly adept at shuttling units among previously prepared launch sites.

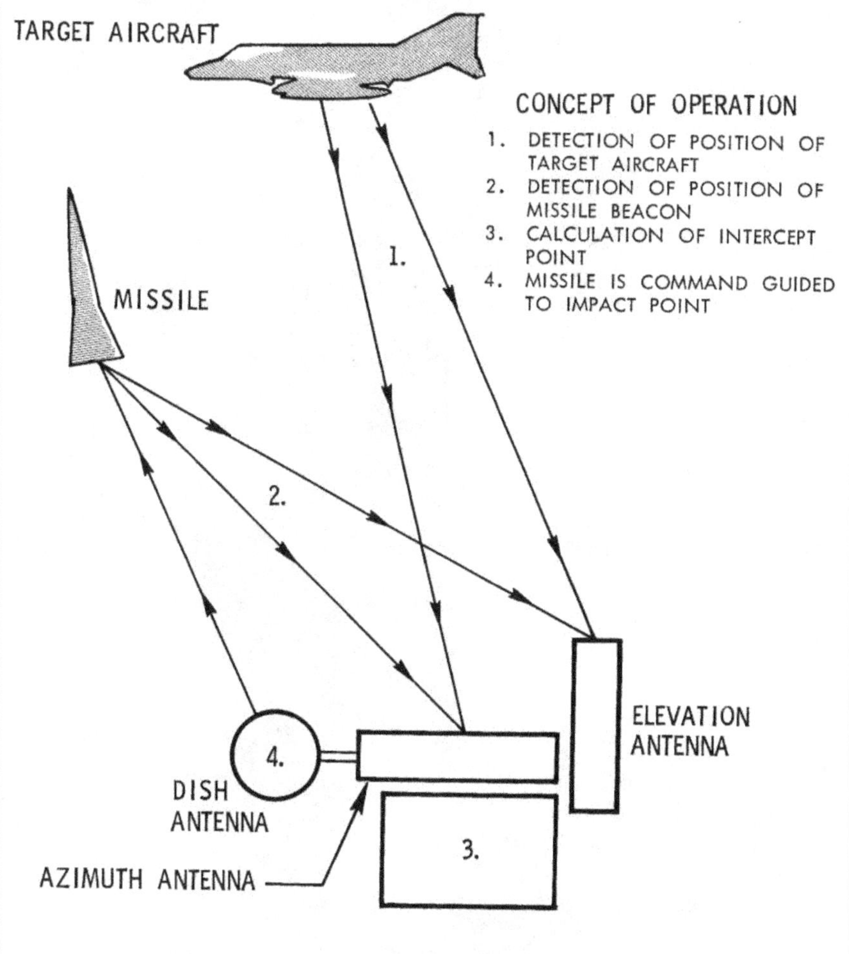

FAN SONG RADAR
AS IT TRACKS BOTH TARGET AND MISSILE

The SAM in its 1965 version had certain exploitable weaknesses other than its inability to engage low-flying aircraft. To provide a sharply defined radar image,

Fan Song operators doubled the pulse repetition frequency some 30 to 40 seconds before launch, and the guidance signal, which could not be delayed more than 4 seconds after launch, confirmed that a missile was on its way. Radar warning gear, installed in American combat planes, could pick up these signals and alert the pilot to make a diving or climbing turn, maneuvers that the missile could not duplicate.

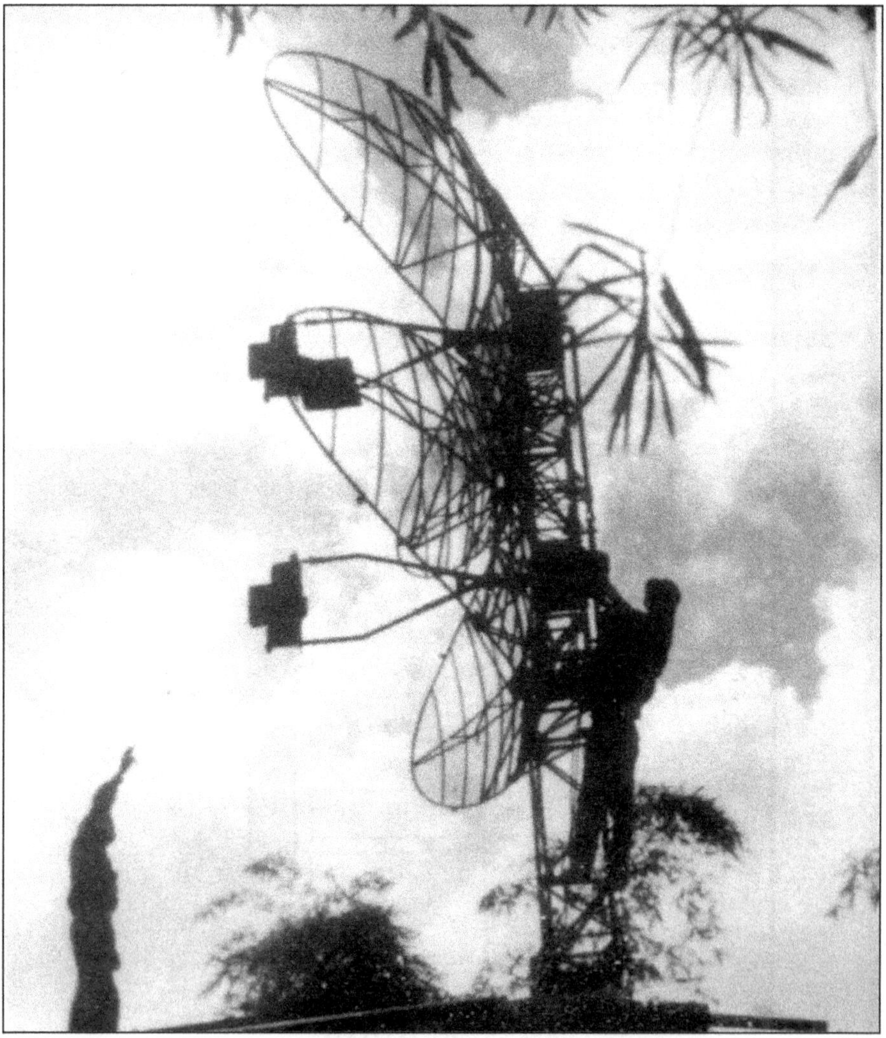

A P-15 (NATO Codename Flat Face) Target Acquisition Radar. Source: People's Liberation Army of Vietnam.

The SAM of 1965 differed, however, from the model in use 7 years later. An important modification was the addition of an optical sighting system, first introduced in 1968, which enabled missile controllers to track aircraft flying as low as 1000 feet. Contained in a box mounted on the horizontal trough antenna, this aiming device was immune to jamming, though dependent on good visibility. Optical guidance required only two electronic signals. One was the guidance command, broadcast from the Fan Song van beginning about 4 seconds after launch. The other was the down link, or transponder signal, which showed whether the missile was following the proper trajectory. Optical guidance gave North Vietnam's defenders an opportunity for ambush. As the attackers approached, a Fan Song radar could begin transmitting to attract the attention of the air crews. Meanwhile, a different SAM site would track the planes optically in order to unleash missiles from an unexpected direction. [5]

Fan Song radar being deployed. Source: Vietnamese People's Liberation Army

In addition, the Fan Song radar itself underwent modification, emerging as a greater menace to low-flying aircraft. In 1972, crews of General Dynamics F-111A's on night missions reported being tracked by Fan Song at altitudes below 500 feet, but they suffered no injury, for the missile warhead did not arm in time to engage a target flying lower than 1000 feet. [6]

The final year of the war saw the North Vietnamese introduce several new pieces of equipment, among them a new surface-to-air missile and a modified Fire Can radar, both of which could have been Chinese inspired. As early as May 1972, Air Force pilots reported seeing a "short, fat, and black" missile with "extremely good guidance and much better maneuverability than the SA-2." [7] The missile, however, did not prove as deadly as the first reports seemed to indicate. On the basis of infrequent sightings during the summer, intelligence analysts concluded that the weapon had essentially the same characteristics as current models of the SA-2. [8]

The other electronic innovation was an I-band radar signal which the Americans at first called T-8209 but later redesignated B-4272 -- Teamwork. This signal emanated from Fan Song radars that apparently had been modified to shift from the usual E-band frequency range to the I-band. By December 1972 the enemy was installing the modified Fire Cans at SAM sites, enabling radar operators to shift from the heavily jammed E and F bands to a less vulnerable set of frequencies. Against the Teamwork signal American electronic warfare officers employed a jamming transmitter designed primarily to deal with airborne intercept radar. [9]

Other Elements of North Vietnam's Air Defenses

To compensate for the ineffectiveness of early SAMs against low-altitude attack, the North Vietnamese turned to conventional antiaircraft guns, surrounding their SAM batteries (and other likely targets) with automatic weapons ranging in size from 12.7-mm machine guns to 57-mm cannon. Russian-designed 85-mm and 100-mm guns also challenged the Americans, but these were fewer in number and less deadly than the lighter weapons. In short, North Vietnam bristled with anti-aircraft guns, aimed by optical sights or radar, that not only covered the low altitudes where SAMs were least effective but also supplemented the missiles in defending against aircraft flying between 3000 and 40, 000 feet.

North Vietnamese 37mm gun crew in a posed picture. Source: People's Liberation Army of Vietnam

Three types of antiaircraft weapons usually mounted optical sights. The lightest was the 12.7-mm machine gun, effective up to 3,000 feet and capable of firing 80 (original text; the correct number is 800) rounds per minute. The heavier 14.5-mm machine gun boasted an effective range of 4,500 feet and a rate of fire of 150 rounds per minute. The 37-mm gun could fire 80 rounds per minute and down a plane flying as high as 5,500 feet. [10]

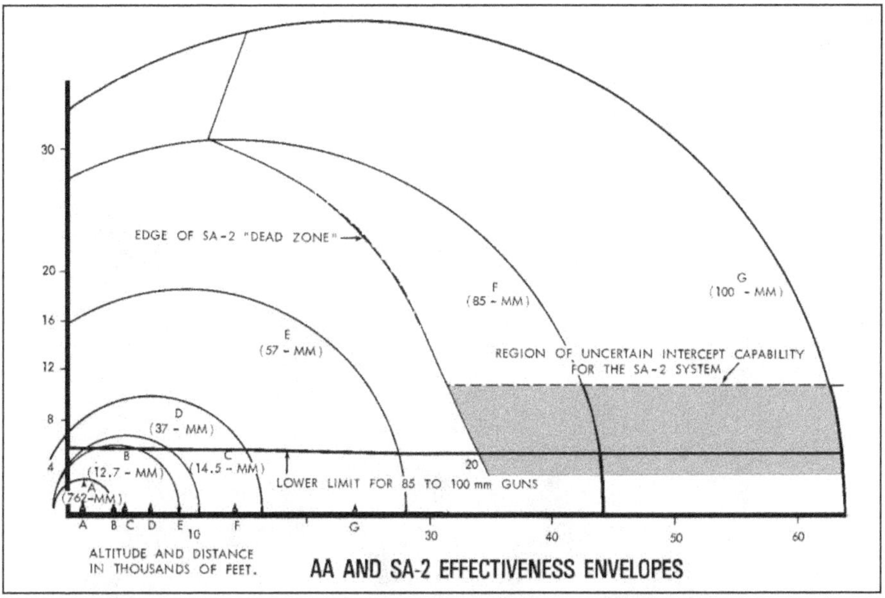

AA AND SA-2 EFFECTIVENESS ENVELOPES

In protecting the SAM sites, North Vietnam's defenders relied heavily on batteries of four to eight 57-mm guns, either towed or self-propelled. A van-mounted Fire Can radar normally controlled these weapons, feeding the data it gathered into a fire control computer, but optical tracking and ranging equipment was also available. Credited with an 80 rounds-per-minute rate of fire, the radar controlled 57-mm gun, could engage a target at nearly 20, 000 feet, Optical aiming, however, reduced the maximum effective range to about 13,000 feet. [11]

The Fire Can radar, also used with the 85-mm and 100-mm antiaircraft guns, evolved from a Russian set that, in turn, was based on an American type produced during World War II. The operator could pick up an aircraft at a range slightly in excess of 50 nautical miles, but the lack of a moving target indicator on the viewing scope complicated the task of tracking fast, low-flying planes. Fire Can was more vulnerable to deliberate electronic interference than Russia's newer gun-laying radars, some of which reached North Vietnam before the fighting ended. [12]

Fire Can (SON-9) Radar. Source: Chinese People's Liberation Army

North Vietnam's heavier antiaircraft guns were deadliest when controlled by radar. The 85-mm weapon had a maximum effective range of 27,500 feet and a rate of fire between 15 and 20 rounds per minute. The 100-mm gun could fire 15 rounds per minute against targets up to 39, 000 feet. [13]

The KS-19 100mm gun was a good gun for its era but the day of heavy anti-aircraft guns like this was passing. Source: U.S. Army

Against aircraft flying so low that neither radar nor optical equipment could track them, the defenders employed barrages from 37-mm and 57-mm antiaircraft guns, machine guns, and even automatic rifles. These weapons did not take aim at the approaching formation but tried instead to throw a wall of fire across the route the attackers would use. According to North Vietnamese Army manuals, a trained rifle platoon could fire a barrage of 1000 rounds within 5 seconds or less. [14]

As American aircraft began attacking from low altitudes to avoid the SAM, the 57 -mm guns, either optically or radar controlled, quickly became the enemy's dangerous weapon. Compilations of American losses from the beginning of Rolling Thunder in March 1965 to the end of that year disclosed that this weapon, together with machine guns and 37-mm guns, accounted for 90 percent of the total. [15]

A third element of North Vietnam's air defenses was the interceptor force. When the F-4C' s of Leopard flight hurtled northward on 24 July, they were protecting the day's strike force against Russian-designed Mikoyan-Gurevich (MiG) fighters. North Vietnamese pilots in MiG-17's had made the first aerial kills of the war in

April 1965, shooting down two Republic F-105D Thunderchiefs near the town of Thanh Hoa. Although subsonic and short-range, the MiG-17 packed three cannon, could carry air-to-air rockets, and was exceptionally maneuverable. By the end of 1965, the more formidable delta-wing MiG-21, super-sonic and armed with both cannon and infra-red homing missiles, was in the hands of North Vietnamese pilots. (Defense Lion Publications notes that, many years after this was written, information from the People's Liberation Air Force of Vietnam claimed that they had a few MiG-21s armed with two 30mm cannon and a lot more armed with two AA-2 Atoll missiles but none with both guns and missiles). As they gained experience, these fliers became increasingly skilled at following instructions from ground controllers to make high-speed, hit-and-run attacks on American fighter-bombers. [16]

North Vietnam's MiG-21 interceptors proved effective at hit-and-run attacks. Source: Vietnamese People's Liberation Air Force.

The Defensive System Evolves

Such were the three components of North Vietnam's air defenses: surface-to-air missiles, antiaircraft guns supplemented by machine guns and even automatic rifles, and modern interceptors. As time passed, the enemy established a communication net that enabled him to use what the commander of an Air Force electronic warfare squadron called "unpredictable operating procedure."

The "classical example" cited by this officer, Col. Morris Shiver, was the use of radars other than Fan Song to track targets for SAM units. Either the battalion's acquisition radar or a nearby early warning set fed data to the SAM fire control center, so that the Fan Song, instead of picking up the American formation at the usual distance of 35 nautical miles, remained in what was called "dummy load, " which left it ready to transmit instantly even though no electrical power was yet reaching the antennas. When the approaching aircraft came within missile range, the Fan Song operator began transmitting in the high pulse repetition frequency, thus avoiding the characteristic doubling of pulse repetition frequency just prior to launch. "This bit of tactical adaptability" Shiver declared, "clearly demonstrated the danger of being complacent where the North Vietnamese are concerned. " [17]

The evolution of North Vietnam's air defenses continued throughout the conflict, During 1971, Defense Intelligence Agency specialists reported signs that the enemy was centralizing control of his SAM battalions at Hanoi's Bac Mai air field, already the site of a command post for interceptor operations. Such a facility, they predicted, could ultimately coordinate the activity of several regional defense centers, each with its own missile and antiaircraft batteries. Yet, when the air war approached its climax in December 1972, the heaviest attacks fell upon Hanoi and Haiphong, and a nation-wide command and control network proved unnecessary. [18]

According to Air Force endorsed intelligence, the number of SA-2 units increased rapidly during the early years of the conflict, then became fairly stable. From seven launch sites at the end of July 1965, the number burgeoned to 20 or 25 battalions in 1967, to 35 or 40 in 1968, and to 45 in 1972. Evidence available to Strategic Air Command planners in December 1972 showed that 26 of the SAM units were located north of the 20th parallel, with 12 of them clustered around Hanoi and 9 around Haiphong. Each battalion had its own Fan Song and target acquisition radars.

Typical SA-2 Site. Source: U.S. Air Force.

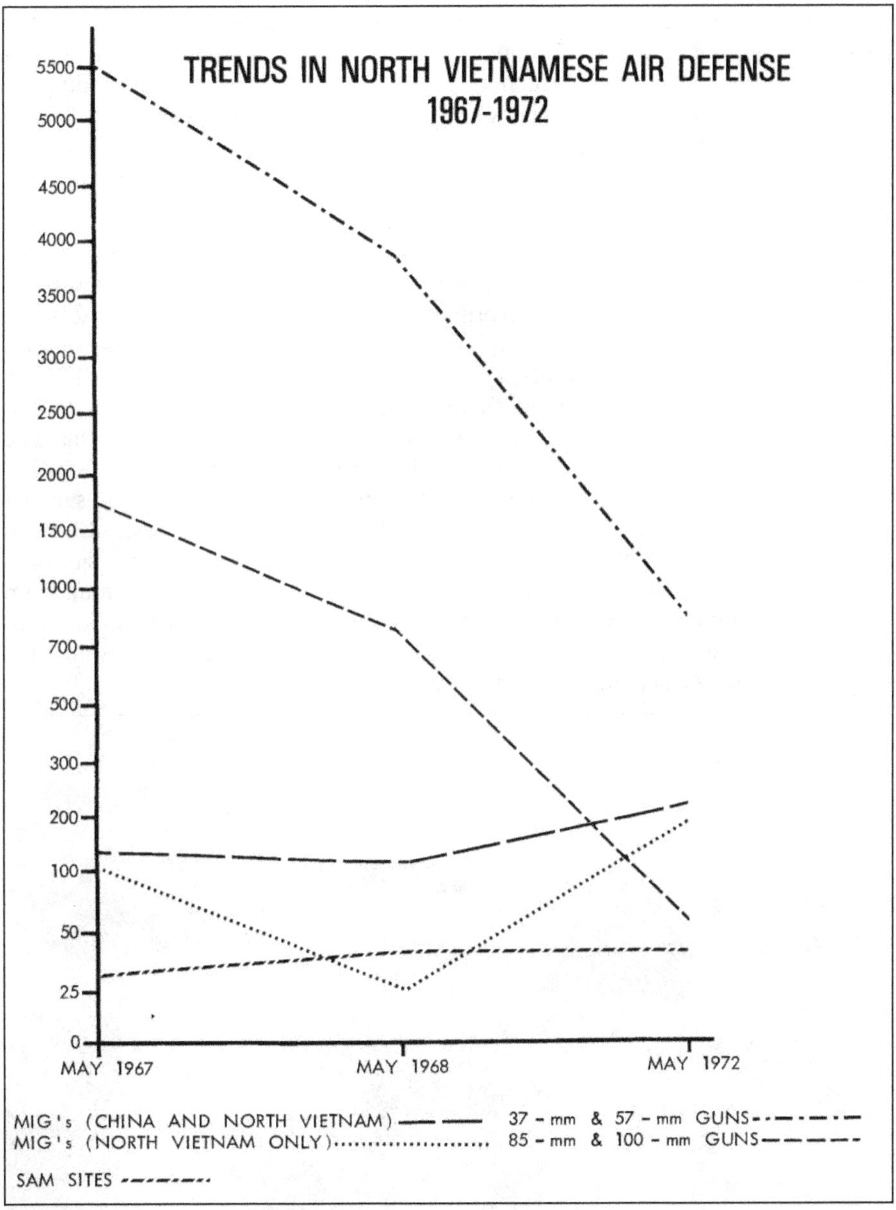

After 1967, intelligence estimates advising Air Force headquarters of the number of North Vietnamese antiaircraft weapons showed a steady decline, as collection technique improved. In the spring of 1967, the Air Force compilation credited North Vietnam with almost 8000 guns, ranging in size from 37-mm to 100-mm. Within a year, a revision had cut this inflated total almost in half, and by the fall of 1969, it had been halved again to approximately 2000 weapons. In the spring of

1972, Air Force headquarters was citing intelligence estimates listing just 884 antiaircraft guns, 824 of them 37-mm or 57-mm and the remainder 85-mm or 100-mm. To some extent, the reduction since 1969 reflected the disbanding of antiaircraft units after Rolling Thunder had come to an end. Meanwhile, Pacific Command intelligence analysts agreed that the number of gunlaying radars, both Fire Can and more modern types, had increased from roughly 30 in 1965 to 49 in 1972.

No such drastic revisions characterized estimates of the number of interceptors available for the defense of North Vietnam. Throughout the conflict, however, the number of MiGs actually in the North fluctuated in response to American air power, which at times forced most of them to retreat beyond the Chinese border. In the spring of 1968, for example, intelligence reported as few as eight MiG-21's and seven MiG-17s flying from North Vietnamese airfields, with the remainder of the force, more than 100 planes, enjoying the security of bases inside China. The aggregate number of fighters carrying North Vietnamese markings varied during 1967 and 1968 from slightly more than 100 to almost 150. The year 1969 saw the total exceed 200, as supersonic MiG-19s and MiG-21s moved south from China to increase from 55 to 126 the number of fighters based in North Vietnam. Air Force intelligence believed that 213 of North Vietnam's 245 MiGs were operating from airfields within the country at the time of the spring 1972 invasion of the South. In December of that year, after almost 9 months of aerial combat, SAC credited the enemy with 144 MiGs on hand for the defense of his homeland. [19]

MiG-17 interceptors. Contrary to many post-war accounts, the MiG-17 was very much of secondary importance. Source: Vietnamese People's Liberation Army

Radar, however, not the number of planes, was the key to the success of the interceptor force. By late 1965, North Vietnamese radar operators had acquired the skill and equipment to control the MiGs from the ground. As early as the summer of 1966, Pacific Command intelligence had conceded that the early warning network could detect an F-105D 15,000 feet above the northeast corner of Thailand, track the plane over Laos, and follow its course over North Vietnam, losing it only when the pilot descended below 3000 feet to take advantage of terrain masking. [20]

When the Boeing B-52 Stratofortresses began systematic night attacks in the Hanoi-Haiphong region during December 1972, SAC planners anticipated a savage reaction by radar-controlled MiGs. This opposition failed to materialize, however, probably because the North Vietnamese pilots were inexperienced in night flying. The principal contribution of the MiGs, aside from an occasional unsuccessful attack, was to help SAM batteries determine the altitude of the bomber stream. [21]

Historical Precedent: Breaking Germany's Defensive Boxes

The defenses confronting Leopard flights over North Vietnam were much like those encountered by British night bomber crews in World War II, the use of surface-to-air missiles being the most significant difference. As early as 1942, however, the German air defense net included not only early warning radar, but radio-directed fighters, and antiaircraft guns, all closely coordinated in air defense sectors or boxes. To penetrate this barrier the Allies developed countermeasures to interfere with radar tracking and radio communication.

The first and simplest of these countermeasures, devised by the British, was window, called chaff by the Americans. Chaff consisted of metallic strips that reflected radar waves and produced strong returns on viewing scopes. Clouds of chaff drifting slowly earthward could mislead radar operators or screen attackers from detection and tracking. During 1943, chaff, became an essential aid to night bombardment. The defenders soon learned to counter it by either switching to radar wave lengths not covered by chaff or by using equipped night fighters with enough endurance to wait for the chaff to disperse.

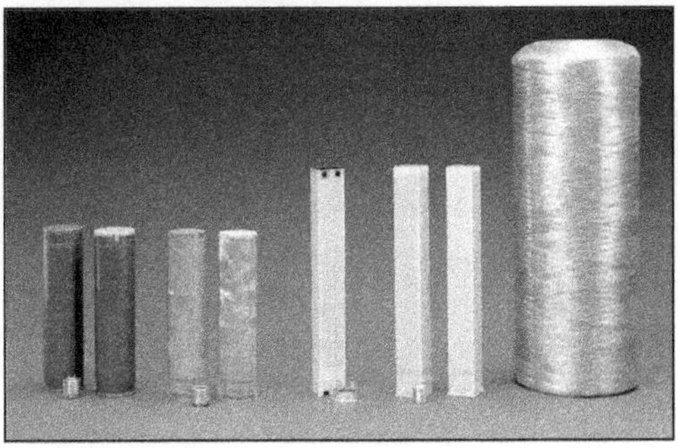

A variety of chaff cartridges produced by Esterline Defense Technologies. Originally, chaff was pre-cut and could only jam predetermined frequencies. These days it is cut to length on demand. Source: Esterline Defense Tech.

British scientists responded with jamming transmitters. One such device, called Cigar, enabled Allied aircraft to broadcast a signal that disrupted radio communication among defending pilots and ground controllers. Another, Mandrel, drowned out the return from early warning radar, causing clutter similar to that produced by chaff. Mandrel stations in the British Isles broadcast for the first time on 5 June 1944, helping to screen the Normandy invasion force.

Operating panel for AN/APT-3, the U.S. version of the British Mandrel jammer. Mandrel was targeted toward the 125MHz German Freya radar. It covered a band of 85-135MHz with a power output of 12 watts. Source: U.S. Air Force.

During the final year of the war, the Allies used Mandrel, Cigar and chaff in combination. The Mandrel ground stations jammed German early warning radar, creating an electronic screen from which a few decoy bombers emerged. These decoys released clouds of chaff to create the illusion of an approaching strike force. The real bomber formation, also dispensing chaff, advanced with the added protection of airborne Mandrel sets, installed in American Boeing B-17s or British Short Stirlings, which blanketed the frequencies used by German ground control intercept radars. The attackers also employed converted bombers carrying airborne cigar which jammed the radio channels relied upon by ground controllers. [22]

Protecting the Strategic Bombers

After World War IL the U.S. Strategic Air Command turned to electronic countermeasures as it studied the problem of penetrating the defenses of the Soviet Union. Electronic warfare training was conducted at Barksdale AFB, Louisiana, where the 376th Bombardment Wing turned out operators for the spot jammers installed in Boeing B-29 and B-50 Superfortresses. These spot jammers were transmitters which concentrated all their available power in a narrow frequency band to create clutter, also called noise, on defensive radar scopes. Accurate electronic intelligence was essential since the operator had to know the exact enemy radar frequencies but until the defending operators received equipment that offered them a choice of frequencies, spot jamming was effective. [23]

Again during the Korean conflict electronic countermeasures demonstrated their wartime value. Spot jamming, inaugurated in April 1951, became standard practice for B-29 night strikes against North Korea. Usually each B-29 carried three spot jammers and a radar receiver to warn that it was being tracked. Sometimes a pulse analyzer, used to identify the tracking radar type replaced one of the jamming transmitters. The spot jammers had different frequency settings, enabling a bomber formation to generate a barrage of electronic noise to blind the various radars that might be encountered. Chaff, which made its Korean debut in September 1952, either reinforced the spot jammers or blanketed frequencies they did not cover. [24]

Locating and neutralizing enemy radar sites was a primary role for the strategic reconnaissance RB-36, RB-47 and RB-52 force. Source: U.S. Air Force.

Equipment developed for World War II performed adequately over Korea, but by the mid-fifties defensive radar had become less susceptible to existing countermeasures. A radar operator could escape the narrow focus of spot jamming by shifting frequencies through a fairly wide range. Barrage jamming, in which the operator distributed power uniformly over a wide frequency band, generated too weak a signal to drown out the return from some of the newer and more powerful radars. A logical solution was to develop sweep jammers that would enable an airborne electronic countermeasures specialist to move a concentrated beam over a fairly wide spectrum, consulting his receiver and pulse analyzer every few seconds to make sure he was transmitting on the exact frequency being used by the enemy. [25]

SAC helped test the newly developed countermeasures gear and installed much of it in the command's newer aircraft. For a time, the 301st and 376th Bombardment Wings, with a total of 90 Boeing EB-47's, were assigned the responsibility during wartime of flying prescribed routes, using chaff and electronic transmitters to confuse the enemy and screen the approach of nuclear-armed bombers. The more modern of SAC's B-52 Stratofortresses carried elaborate jamming equipment to improve chances of survival during low-altitude solo penetrations of radar-controlled defenses. [26]

Countermeasures for Tactical Aircraft

Although Tactical Air Command (TAC) also was interested in electronic warfare, development of the necessary equipment did not receive an overriding priority. Work in this field went ahead, however, despite the difficulty of finding space for countermeasures devices in compact tactical aircraft and a chronic shortage of funds for research and development. By 1957, TAC had radar warning receivers and chaff dispensers available for its most modern fighter-bomber, the North American F-l00D, and for the British-designed Martin B-57 light bomber. The new McDonnell RF-101 reconnaissance plane carried just the warning equipment.

Among the countermeasures devices under consideration in 1957 was a jamming pod to be mounted under the wings of fighter-bombers instead of inside the already crowded air frame. The pod could be removed for strikes where radar-controlled weapons were not anticipated and replaced by an equal weight of munitions or fuel. This self-protection pod, as it came to be called, contained only two transmitters, so coverage was limited to a narrow frequency range. Consequently, to compensate, strike planners had to have precise intelligence on enemy radar. [27]

QRC-160 pod being loaded on to a F-105. Source: U.S. Air Force.

By the end of 1963, the Air Force was testing a family of five QRC-160 self-protection pods, one of them designed to jam Fan Song. Fitted to pylons on the F-100D and the newer Republic F-105, the pods weighed 79 to 90 kilograms (175 and 200 pounds, measures 250 centimeters (100 inches) in length by 25 centimeters (10 inches) in diameter, and contained a propeller- driven generator to provide electric power. Similar pods were planned for the RF-101, the McDonnell F-4 and RF-4, and the General Dynamics fighter-bomber that ultimately became the F-111. [28]

The most extensively equipped electronic warfare planes available to TAC in the early 1960's were modifications of the Douglas B-66 Destroyer, a twin-jet, shoulder-wing light bomber developed for the Navy and then manufactured for both the Navy and the Air Force. One of the two variants was used principally for electronic reconnaissance and the other exclusively for jamming. [29]

The RB-66C was the electronic intelligence part of the team. Source: U.S. Air Force

The electronic reconnaissance model was the RB-66C, created by replacing the camera compartment and photoflash bomb bay of a standard RB-66 with a pressurized capsule that housed four electronic warfare specialists and a variety of gear. This equipment, some of it developed for SAC strategic bombers, included radar receivers, direction finders, pulse analyzers, chaff dispensers, and jamming transmitters. During the early 1960s, RB-66C's deployed from Shaw AFB, South Carolina, to Europe for training that included flights along the perimeter of the Iron Curtain to ferret out and record transmitting frequencies and location of East German radars. [30]

The RB-66B (later modified to EB-66Es shown here) were the jamming component of the team: Source: U.S. Air Force.

The jamming version was an RB-66B, minus its cameras and related equipment, but retaining its permanently installed radar warning receiver. It carried a special bomb bay pallet, called a cradle, that accommodated both counter-measures transmitters and chaff dispensers. A tail cone containing electronic countermeasures gear replaced the usual gun turret. An array of 23 jamming devices enabled the aircraft's one electronic warfare officer to lay down an effective noise barrage over a wide frequency spectrum. [31]

Like the RB-66C's, those B models fitted out for electronic warfare flew numerous peacetime training missions, testing their jammers against American radar. Out of this experience, a tactical doctrine evolved that called for countermeasures-carrying RB-66B's to accompany bomb-laden B-66's, jamming enemy radar throughout the attack. By 1965, however, the B-66 was no longer a first-line light bomber, and the practice was never followed in Southeast Asia. [32]

Prior to the Vietnam war. TAC was confident that it was developing and gradually acquiring countermeasures gear that would disrupt modern radar-controlled defenses. The watchword at the time was "quality rather than quantity." [33]

Once the Southeast Asia fighting began, however, North Vietnam's defenses improved with stunning speed, and TAC found itself struggling to keep pace.

II. THE EB-66 AND STAND-OFF JAMMING

At the time of the first Rolling Thunder attacks in the Spring of 1965 neither RB-66Cs nor countermeasures-equipped RB-66Bs had arrived in Southeast Asia. Although TAC had some 65 RB-66s, 58 of them ready for combat, not all were modified for electronic warfare. And with this small force, the command had to fulfill commitments to U.S. Air Force Europe (USAFE), conduct training, take part in joint exercises, and meet the needs of the Vietnam war. To ease the shortage of aircraft, and also to obtain scarce parts, TAC turned to the 3 dozen retired B-66's stored at Davis-Monthan AFB, Arizona.

The RB-66Bs were in short supply right from the start. Source: U.S. Air Force

Despite other demands, the command did send six EB-66Cs to Southeast Asia. These began flying from Takhli Air Base, Thailand, in May 1965. This deployment exhausted the pool of C models until September, when TAC was able to increase the number of EB-66Cs at Takhli to nine, but it was unable to dispatch any B models until October, when five of them arrived from the United States. Eight more EB-66Bs, made available from USAFE, reached Thailand in May 1966.

At this time, Air Force statisticians used the designation RB-66 whether the plane carried cameras, infra-red detectors, or radar detection and jamming equipment. A distinction was later made between RB-66Cs and Bs, though the latter might perform reconnaissance rather than countermeasures duty. The RB-66Bs remained lumped together until the spring of 1966, when the prefix E was assigned to all versions of the B-66 engaging in electronic warfare. Henceforth, the E prefix will be used in this narrative for all electronic warfare variants -- EB-66Bs, Cs and after August 1967 EB-66Es -- that served in Southeast Asia.

The planes were grouped according to type. The Cs served with the 41st Tactical Reconnaissance Squadron, redesignated the 41st Tactical Electronic Warfare Squadron in September 1966. The Bs joined the 6460th Tactical Reconnaissance Squadron, redesignated the 42d Tactical Electronic Warfare Squadron in September 1966. Both units became components of the Takhli-based 355th Tactical Fighter Wing in August 1967. The maximum combined strength of the two squadrons was 37 aircraft in May 1968. [1]

Early Operations

The EB-66C's entered combat in May 1965. Their principal mission was to gather intelligence on enemy radar, concentrating on Fan Song, whose signal might indicate that a North Vietnamese SAM site was ready for action. On 23 July, one of these planes intercepted a Fan Song transmission. The following day, an EB-66C picked up another such signal confirming the site's activity and just seconds later a SAM destroyed Leopard 02. [2]

Since the EB-66C's focused upon intelligence collection and SAM warning, Marine Corps and Navy planes did most of the radar jamming until the first EB-66B's arrived at Takhli in October 1965. Each of the three aircraft carriers operating in the Tonkin Gulf during the spring of 1965 carried a 4-plane electronic warfare detachment. The usual aircraft assigned to these units was the Douglas EA-3B Skywarrior, based on the same light bomber design as the EB-66. Some of the detachments also flew the EA-1B, a modified Douglas Sky raider single-engine, piston-powered attack plane. A few Navy pilots, and the Marines based at Da Nang in South Vietnam, flew the Douglas EF-10B Skyknight, a twin-jet fighter first flown in the late 1940s and later converted for electronic warfare. [3]

During the spring and summer of 1965, Marine EF-l0B Skyknights usually laid down the jamming barrages for Air Force strikes against the North. This modified Skyknight had two spot jamming transmitters installed in the fuselage. The plane could carry an external pod containing a combination of four noise and deception jammers, the latter designed to deceive enemy radar operators by broadcasting a false return. The EF-l0B also had external fittings to accommodate an external fuel tank and a chaff dispenser.

Because the Skynight lacked aerial refueling equipment, those EF-10Bs supporting air attacks in the Red River delta had to carry a 300-gallon auxiliary fuel tank in order to spend even 50 minutes orbiting over the Tonkin Gulf. Despite the extra tank, Marine fliers sometimes shut down one engine to conserve fuel while descending from their usual operating altitude of 30,000 to 35,000 feet during the return flight to Da Nang. Furthermore, the weight of the additional fuel, plus the drag created by the external container, reduced the plane's already marginal performance. So sluggish was the rate of climb of a fully loaded EF-10B that the planes took off toward the sea to avoid small-arms fire from Viet Cong guerrillas concealed along the airfield's inland perimeter.

The EF-10B Skynight filled the gap while the EB-66 force was being assembled. The problem was there were only about 10 EF-10Bs available and their capabilities were severely limited. Source: U.S. Navy

The Marines enjoyed some success when three or four EF-10Bs, orbiting over the Tonkin Gulf, dropped chaff and transmitted jamming signals over a frequency range broad enough to embrace Fire Can gun control early warning, and ground control intercept radars. Captain John L. Pycior, USMC, an EF-10B pilot, was confident that Skyknights, transmitting from the comparative safety of the gulf, could disrupt radars along the coast line, but he conceded that the planes were only marginally effective against targets more than 16 nautical miles inland.

Until the SAM threat became inordinate, Skyknights ventured inland to jam Fire Can and sometimes Fan Song sets. In order to blanket Fire Can, three or four EF-10Bs had to fly a circular orbit, with the target in the center, and transmit from an altitude of 20,000 to 25,000 feet. The Marines usually challenged Fan Song during night interdiction missions. The electronic warfare officers on board listened for track-while-scan and guidance signals, radioed SAM warnings to the light bomber they were escorting, and then tried to jam the Fan Song beam.

Because it lacked the jamming power of even an EB-66C and the speed and maneuverability of a fighter-bomber the EF-10B was a poor risk over SAM-infested areas of North Vietnam. By late 1965, the proliferation of missile sites and the increased skill of SAM crews had forced the Skyknights to orbit too far from most inland targets for effective radar jamming. From April until October, however, the Marine plane had been an acceptable stand- in for the Air Force EB-66B. [4]

The EB- 66 Enters the Lists

The EB- 66B and EB-66C enjoyed several advantages over the Marine EF-10B Skyknight. The Air Force planes could refuel from suitably equipped aerial tankers. They carried more jamming equipment -- nine transmitters in the C model and as many as 23 electronic devices and chaff dispensers in the EB-66B -- compared to the Skyknight's six less powerful jammers and single chaff dispenser. The Skyknight, moreover, did not have the extensive intelligence gathering equipment found in the EB-66C. [5]

As the RB-66C force built up, the aircraft took on an increasing level of the combat support function – and an increasing level of risk. Source: U.S. Air Force

The EB-66C's based at Takhli took part in the retaliatory attacks that followed the destruction of Leopard 02 and in other strikes against the North. On a typical mission, a pair of EB-66Cs took off from Takhli topped off their fuel tanks from a Boeing KC-135 Stratotanker, and rendezvoused with the strike force. The two aircraft accompanied the fighter-bombers to the vicinity of the target, then entered an elliptical orbit at 25,000 to 30,000 feet, beyond reach of 37-mm or 57-mm antiaircraft guns. The electronic warfare officers jammed Fire Can radars, while listening for Fan Song signals. If the crew picked up the doubling of the Fan Song pulse repetition frequency or the SAM guidance signal, the chief electronic warfare officer alerted the strike force by radio and joined the other countermeasures operators in jamming the Fan Song tracking beam. [6]

The EB-66Bs made their combat debut in October 1965 and soon demonstrated that the very jamming power that enabled them to close with the target could also be a defensive liability, for the electronic noise emanating from the plane interfered with its radar warning equipment. As a result, the EB-66B and C had to work together. The broader and stronger jamming barrage laid down by the Bs afforded better protection against the SAMs, allowing the C models to remain well beyond the 17-nautical-mile range of the Guideline missile. In practice, therefore, while one or two B models orbited within 15 nautical miles of the target, an EB-66C remained safely beyond missile range, providing SAM warning and ensuring that the jamming barrage blanketed those frequencies the enemy was using. [7]

The air-to-air refueling capability of the RB-66 was tactically very useful. Unfortunately, the aircraft used the old probe-and-drogue system which restricted the availability of tankers. Here an RB-66 refuels from a KB-50. Source: U.S. Air Force.

Those officers who planned EB-66 jamming missions had to take into account the fact that electronic noise did not radiate from the fixed antennas in a uniform, concentric pattern. Indeed, antenna location caused the jamming coverage to resemble a sort of Rorschach butterfly, the plane at its center and the strongest signals radiating perpendicular to the flight path. For this reason, planners tried to assign the planes in pairs, arranging the orbit so that one of them was always broadside to the hostile radar. [8]

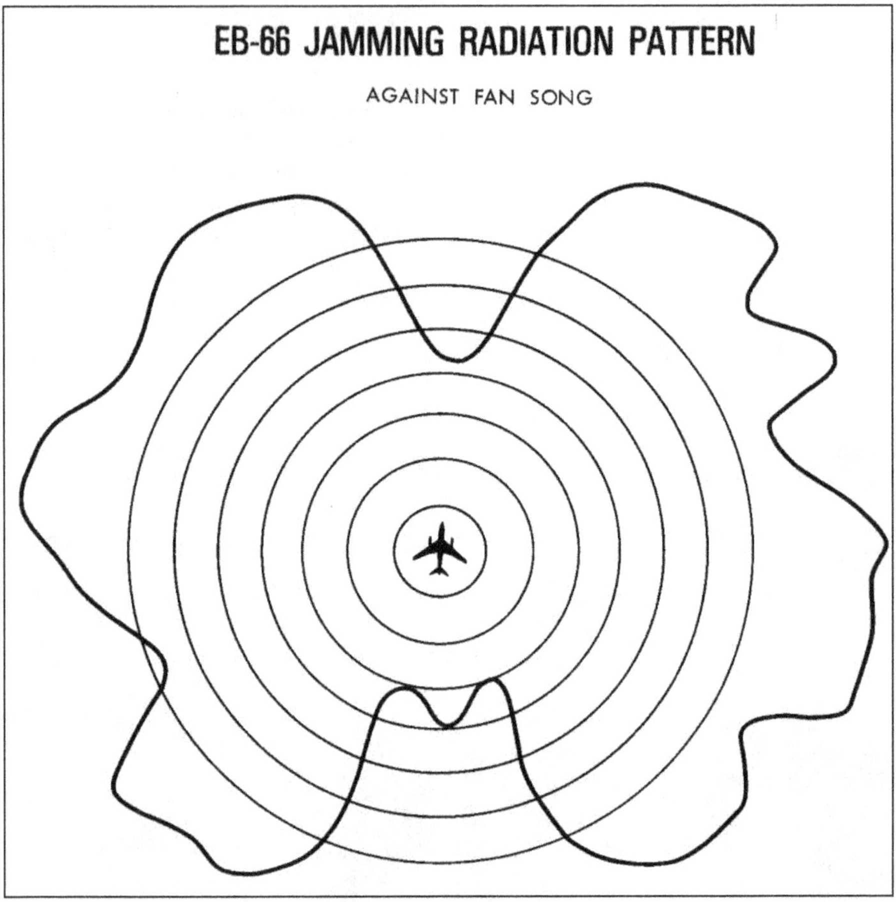

Action and Reaction, 1966 - 1968

Throughout the Rolling Thunder campaign, EB-66 jamming involved compromise between effective coverage and aircraft survival. Although the effectiveness of electronic jammers decreased as distance to the target radar increased, distance protected the planes from hostile fire. Furthermore, the noise barrage gave the best protection when the attacking fighter-bombers were between jamming orbit and target, but the enemy could shift his weapons to prevent the EB-66s from assuming this ideal station. [9]

As late as February 1966, the planes were reasonably safe if they flew too high for light antiaircraft guns and avoided the missile defenses that ringed Hanoi,

Haiphong, and other nearby targets. To support strikes in this heavily defended region, they flew orbits over the Tonkin Gulf and inland above the mountains northwest of the Red River delta. Together, these two stations provided excellent coverage, for they bracketed the area where the North Vietnamese had concentrated their radar-controlled defenses.

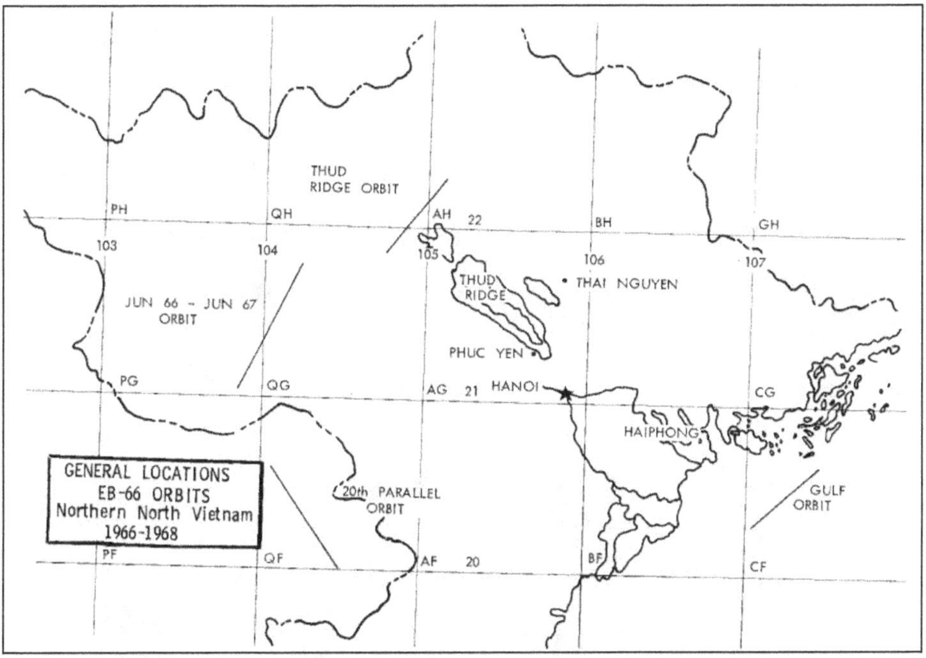

For Air Force fighter-bomber pilots based in Thailand, the inland orbit was the more valuable. In order to find concealment from enemy radar, they hugged a ridge line that pointed southeastward from the barren highlands toward the Hanoi-Haiphong area. This geographic feature was Thud Ridge, so named by the F-105D pilots, who referred to their Thunderchiefs as Thuds. When the Thailand-based aircraft began attacking the delta, EB-66's manned an orbit from which they could transmit directly along Thud Ridge, keeping the strike force between jamming source and target throughout approach and withdrawal. [10]

The freedom of operation enjoyed by the EB-66's (except over the Red River delta) came to an abrupt end in February 1966, when a SAM downed an EB-66C near the town of Vinh, some 140 nautical miles south of Hanoi. The action began when the aircraft crew picked up a weak Fan Song signal and promptly commenced jamming. Next came the pre-launch surge in the Fan Song pulse repetition frequency, which persisted despite continued jamming and an evasive turn. The telltale guidance signal alerted the Americans that a missile was on its way, but before they could maneuver to safety, the warhead exploded, crippling the plane. The crew parachuted into the Gulf of Tonkin, where Navy helicopters rescued all but one of the six persons on board. [11]

The destruction of this aircraft marked the beginning of a southward and westward extension of North Vietnam's SAM defenses. The appearance of new missile sites forced the EB-66's to retreat, though on rare occasions the planes did challenge SAM batteries. In October 1966, for example, an EB-66C spent a quarter of an hour cruising above an area defended by the missiles, trying unsuccessfully to lure the enemy into using a Fan Song transmitter, so that an F-105 cruising nearby could attack with anti-radiation weapons. [12]

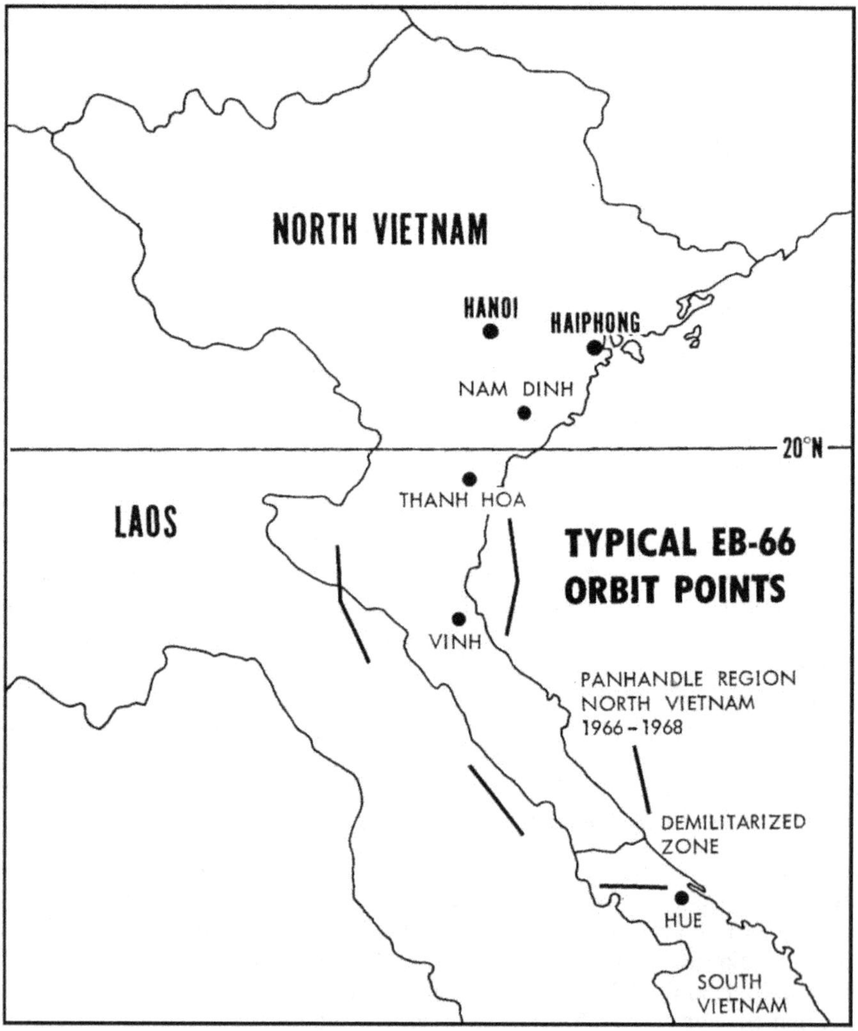

SAMs first appeared in northwestern North Vietnam in mid-1966. This shift forced the EB-66s to move south and west from the original Thud Ridge orbit, increasing both the distance to Hanoi-Haiphong and the angle formed by the jamming source, the target area, and the course generally flown by Thailand-based fighter-bombers. Since the EB-66's had apparently moved to safer skies, Seventh Air Force withdrew fighter cover from the inland orbit. The enemy, taking advantage of this decision, sent MIG interceptors to harass the countermeasures aircraft. Pressure from these enemy interceptors and from SAMs forced the EB-66s, by July 1967, to retreat to new orbits in the vicinity of the 20th parallel, so far to the southwest that noise jamming was ineffectual against Fire Can and Fan Song radars in the Red River delta area. Fortunately, the introduction late in 1966 of jamming pods for individual fighter-bombers and reconnaissance planes offset the effect of withdrawing the EB-66s from Thud Ridge.

Strike forces came to rely on the pods to frustrate gunlaying and missile-control radars guarding vital installations in the delta. This permitted the EB-66s to concentrate on early warning, ground control intercept, and acquisition radars that could be jammed from a more southerly orbit. [13]

An EB-66 at Takhli RTAFB. There are apocryphal and never-confirmed claims that an EB-66 killed a MiG by jamming its radar altimeter, causing it to fly into the ground. This seems unlikely. Source: U.S. Air Force.

For roughly 5 months, the combination of MiGs and SAMs kept the EB-66s tied to the 20th parallel. Then the installation of new radios, designed to ensure reception of MiG warnings even though the jammers were functioning, enabled the EB-66s to advance their inland orbit northward of the 21st parallel. This adjustment, in October 1967, coincided with a contraction of the SAM defenses as the enemy reinforced the Red River delta against intensified air attack. Taking advantage of this North Vietnamese redeployment, the Seventh Air Force in November permitted the EB-66s to orbit above the northwestern extremity of Thud Ridge, but only with fighter cover.

On the second day that escorted EB-66s manned the Thud Ridge station, North Vietnamese MiGs appeared nearby. Seventh Air Force yielded before this show of force by shifting the inland orbit south of the 21st parallel. This move had little impact on Rolling Thunder since a position over Thud Ridge was not essential for the kind of jamming the EB-66s now performed. Unfortunately, this retreat did not placate the enemy whose MiGs shot down an EB-66C on 14 January 1968. Three members of the crew were rescued, but the other four remained prisoners of the North Vietnamese until March 1973.

EB-66s at their base in Thailand. These aircraft and their crews were always small in number and in high demand. Source: U.S. Air Force

Seventh Air Force reacted to this loss by forbidding EB-66s to fly over North Vietnam and by maintaining a barrier patrol of F-4 fighters to screen them from air attack. These policies remained in effect until bombing in the Red River delta ended on 1 April 1968. [14]

EB-66 jamming tactics changed very little while the 1 April bombing restrictions were in effect. Until 31 October 1968, when President Johnson halted the bombing of North Vietnam, the electronic warfare aircraft supported attacks on targets south of the 20th parallel, using orbits over Laos and the Tonkin Gulf. During one typical mission, an EB-66E circled above Laos while another over the gulf jammed radars, and F-105D' s attacked targets in North Vietnam's panhandle. On this occasion electronic intelligence indicated that the EB- 66Es, improved versions of the EB-66B's, had been so effective that the enemy received no radar warning until the attacking aircraft were within 10 to 30 nautical miles of their assigned targets. [15]

Problems and Improvements

The EB-66 had certain inherent weaknesses, most of which stemmed from the fact that the plane was not originally designed for the job it was now required to do. The aircraft engineers who modified the basic RB- 66 for electronic warfare had increased its weight with no corresponding increase in power. As a result, the plane performed sluggishly and in Thailand's hot and humid climate clung tenaciously to the runway during takeoff. In order to reduce the long run needed to coax fully loaded EB-66's into the air, the planes took off with fuel tanks partially full and topped off from aerial tankers. Even so, commented EB- 66 veteran Col. Ian D. Rothwell, the failure of one engine during takeoff meant that "a crash was inevitable," unless the landing gear was retracted and the indicated airspeed was at least 180 knots. [16]

The overworked Allison engines lapped up fuel at a disturbing rate. Luckily, the EB-66's could refuel in flight, using the probe-and-drogue method, inserting the plane's fuel intake into a funnel-shaped receiver at the end of a hose trailing from the tanker. The Air Force F-4s and F-105s flying from Thailand employed the flying boom method, however, in which an operator on board the tanker maneuvered the fuel-carrying pipe into a receptacle in the fuselage of the plane. Most of the tankers stationed over Laos, therefore, mounted the flying boom for their main job of refueling the strike forces, and sometimes an EB-66 on the inland orbit, unable to find a tanker in an emergency had to cut short its mission. But, since many Navy planes used the probe and drogue, an EB-66 running out of fuel over the Tonkin Gulf usually could find a compatible tanker. [17]

The Air Force tried to improve jamming effectiveness, by converting EB-66Bs to E models, the first of which reached Thailand in August 1967. Although this latest variant had 21 jamming devices, two fewer than the B version, its transmitters were tunable, enabling the electronic warfare officer to change frequency during flight and jam different types of radar. In contrast, the EB-66B carried only one adjustable transmitter, which limited choice to three predetermined frequencies. [18]

Mission planners soon devised jamming packages -- instructions telling electronic warfare officers what frequencies to jam, when to transmit, and when to release chaff. These provided adequate countermeasures for the kind of mission the EB-66's were supporting. To obtain the best possible coverage from the package, the aircraft flew a standardized orbit designed for a particular task, such as protection of reconnaissance drones or B-52 bombers. Although individual electronic warfare officers might argue that this standardization told the enemy what sort of mission to expect, the Air Force Electronic Warfare Center maintained that the practice "provided considerable flexibility, while simplifying mission planning and coordination . . . ." [19]

An important aircraft modification was the installation of steerable antennas in the EB-66Cs. This change, begun in the spring of 1968, enabled electronic warfare officers to focus a plane's jamming energy against a specific radar transmitter. The E model never carried this device, probably because the modification would have required the further installation of direction finding equipment to tell the operators where to aim the new antenna. [20]

The Effectiveness of the EB-66

Col. Arthur D. Thomas, who served in Southeast Asia with the 460th Tactical Reconnaissance Wing, reported in October 1966 that the EB-66 was doing "an outstanding job of stand-off jamming." [21] Like most other aspects of the electronic warfare effort, however, the effectiveness of this plane could not be evaluated in terms of missions flown and fighter-bombers lost. Despite the absence of valid supporting statistics, the colonel's judgment is important. He was there, saw the plane in action, and he concluded that it was performing well.

Nevertheless, by 1968 during the latter months of Rolling Thunder the EB-66 did only what the enemy allowed it to do. In late 1966 when Colonel Thomas returned from Southeast Asia, the North Vietnamese were already exerting the pressure that forced the planes into orbits from which they could not jam the Fan Song and Fire Can radars guarding the most heavily defended targets in the Red River delta. Fortunately, the arrival of the self-protection pod offered an alternate means of jamming so that the inadequacy of the EB-66 was not critical and the aircraft could disregard Fan Song and Fire Can and concentrate on early warning, ground control intercept, and target acquisition radars that lay within its jamming range.

The inability of the EB-66 to survive in daylight skies over North Vietnam doomed a plan to use the plane in support of the F-4s protecting American strike forces from MiG interceptors. The EB-66's tried during 1967 to jam the MiG identification equipment relied upon by North Vietnam's ground controllers, but the closest orbit was some 75 nautical miles from the aerial battlefield, too far for an effectual jamming signal. Once again enemy defenses had frustrated the EB-66. [22]

III. WILD WEASEL BARES ITS FANGS

President Johnson, kept tight rein on the air war against the North, but he nevertheless agreed to avenge the loss of Leopard 02. Consequently, on 27 July 1965) a force of 46 F-105 Thunderchiefs, escorted by 12 F-4Cs and 8 Lockheed F-104 Starfighters, attacked two SAM sites and related barracks some 25 nautical miles west of Hanoi. Three of the handful of EB-66C's then in Southeast Asia supported these strikes, providing SAM warning. They also joined six Marine EF-10B's in jamming enemy radar. This day's combat forcefully demonstrated the difficulty in locating and destroying a SAM complex.

The first successful use of the SA-2 missile lured additional U.S. aircraft into a deadly trap. Source: People's Liberation Army of Vietnam

Taking advantage of the SA-2's inability to engage low-flying aircraft, the Thunderchiefs attacked at altitudes between 50 and 100 feet, only to be scourged by fire from automatic weapons. Enemy gunners shot down four F-105Ds, and only one of the pilots was rescued. One Thunderchief sustained flak damage that forced the pilot to break off the action and limp toward home, escorted by another F-105D. After crossing the Thailand border, the pilot of the crippled plane asked his escort to come alongside to determine the extent of the battle damage. As the other aircraft drew near, the battered Thunderchief suddenly pitched upward, colliding with the second plane. Only one of the pilots succeeded in ejecting from the flaming wreckage and his parachute failed to open, so both men perished.

This attempt to punish the enemy for downing Leopard 02, with its two-man crew, cost six aircraft destroyed and five men killed or captured, victims not of SAMs but of light anti-aircraft guns little different from those used in World War II. Aerial photographs taken after the strike showed that one of the missile sites might have been a decoy built to lure the fighter-bombers within range of the automatic weapons. [1] (Defense Lion Publications notes that, long after this was written, the Vietnamese revealed that they had known the attack was coming and which sites would be attacked. They therefore replaced the missiles at the sites in question with dummies and stacked their anti-aircraft defenses around them. The F-105s had flown into a carefully-planned and skillfully-executed trap.)

Within 2 weeks of this fruitless counter thrust, the SAM struck again. On the night of 11 - 12 August 1965, the pilots of two Navy McDonnell Douglas A-4E Skyhawks, on nighttime armed reconnaissance about 50 nautical miles south of Hanoi, saw 2 lights rising toward them through the clouds. Soon realizing the glow was burning SAM propellant, they attempted to escape, but before they could dive to safety, both warheads exploded, downing 1 of the planes and punching some 50 holes in the scorched underside of the survivor.

Navy carrier pilots reacted to the destruction of the Skyhawk with the same low-altitude tactics that Air Force fliers had recently used, and with equally disastrous results. The Navy squadrons not only failed to locate even one SA-2 launcher but lost two pilots and five planes to North Vietnamese antiaircraft fire. "It was," wrote Vice Adm. Malcolm W. Cagle, "truly a black Friday the 13th for TF-77." [2]

The SA-2 sites were very mobile and by the time information on their location had reached the strike forces, they would be long gone. Source: People's Liberation Army of Vietnam

Early in August, Air Force units in Thailand began keeping a few fighter-bombers on alert, fueled and armed to attack newly discovered SAM sites. Reaction time was too slow, however, for the force depended upon photo reconnaissance to pinpoint any SAM launcher that might reveal its general location by firing upon an American formation. To process the film, interpret the pictures, and dispatch a strike took from 6 to 8 hours. During this time, the missile unit could move to some previously prepared site, perhaps leaving behind several of the automatic weapons (that had already proved so deadly against low-flying aircraft) to greet the strikers. As a result, the alert force was disbanded after a few frustrating weeks. [3]

Air Force planners next tried to use radio-controlled drones to trick enemy radars into transmitting so that RB-66Cs or EA-3Bs could locate the Fan Song radars and the SAM battalions they served. The first attempt to use this technique failed to trigger enemy radar. Then, on 31 August 1965, a modified Lockheed C-130 transport launched a second pair of drones over the Gulf of Tonkin, off Da Nang. North Vietnamese radar reacted as the pilotless craft approached Hanoi, and three fixes were obtained, each accurate to within 2 nautical miles. An area near the town of Piu Tho, about 40 nautical miles northwest of Hanoi seemed worth attacking, but the 4 F-105Ds dispatched there failed to find any trace of a SAM site. When fuel ran low, the planes attacked an alternate target, a wooden bridge. One of the planes was lost to antiaircraft fire, though the pilot was rescued. [4]

A classic photograph of a fully-loaded F-105D. This kind of load was very rarely used in combat since, configured this way, the F-105's take-off run was almost equal to its tactical radius. Source: U.S. Air Force.

The use of drones did decrease reaction time, but not enough to make a significant difference. The fighter-bombers could now thunder down the runway just 3 hours after a Fan Song site was located, rather than waiting 6 to 8 hours for photo reconnaissance data. The drones, however, were less precise than aerial photography, so the attackers, after arriving at an indicated position, had to search an area 4 nautical miles in diameter to pinpoint the exact site. Moreover, this dangerous search for a camouflaged and heavily defended missile complex might well be fruitless, since even 3 hours was sufficient time for a SAM battalion to pack up and thread its way over narrow roads to a new location. [5]

What was needed was an airplane that could lead a flight of fighter-bombers into North Vietnam, detect and locate Fan Song transmitters, then direct attacks against the SAM battalions. The Navy already had a few such planes -- Grumman A-6s and those McDonnell Douglas A-4Cs and A-4Es with radar warning and direction finding gear. On 31 October 1965, one of the specially equipped A-4E Skyhawks took off from Takhli to lead eight Air Force Thunderchiefs on an armed reconnaissance of probable North Vietnamese missile sites. The Navy plane located a Fan Song radar, dropped bombs to mark it, and summoned the F-105Ds, which blasted the launchers and fire control equipment. Ironically, the Skyhawk responsible for this destructive strike fell victim to antiaircraft fire, and attempts to rescue the pilot failed. [6]

Toward a Hunter-Killer Team

The F-100F formed the basis for the first Wild Weasel aircraft. Source: U.S. Air Force.

In the United States, meanwhile, both the Navy and Air Force were seeking ways to defeat the SAM. Air Force participation had begun in August 1965, when Gen. John P. McConnell, the Chief of Staff, directed that the threat from both SAMs and antiaircraft guns be analyzed and countermeasures recommended. By early autumn, the Air Force effort was showing results. One type of countermeasure, Wild Weasel I, was undergoing tests at Eglin AFB, Florida. This ferociously named aircraft was an F-100F, the two-seat trainer version of the North American

Super Sabre tactical fighter especially modified to hunt SAM sites. Wild Weasel's crew consisted of a pilot and, seated behind him, an electronic warfare officer with radar homing and warning equipment to locate Fan Song and other transmitters.

Wild Weasel prototypes were equipped with a panoramic scan receiver and a vector homing and warning set to detect and locate enemy radars. Both devices obtained a bearing to the transmitter by comparing the strength of radar signals picked up by antennas installed at various places on the aircraft. These signals appeared as lines of light on viewing scopes mounted in the rear cockpit, and the electronic warfare officer interpreted their characteristics to determine direction to the transmitter and its pulse repetition frequency. A scope located on the pilot's instrument panel duplicated the information available to the electronic warfare officer.

The vector scope picture, plus a flashing light and a chirping noise in the headsets, told the crew that hostile radar was tracking their plane. Another radar receiver picked up the SAM guidance signal and used lights (later supplemented by a buzzing tone in the headset) to signal that a launch was imminent. The pilot could then look for the approaching missile and maneuver to elude it. [7]

Between 11 October and 18 November 1965, four Wild Weasel I prototypes flew test missions against Air Force radars similar to Fan Song. During these experiments, the electronic warfare officer usually made his first contact by means of the panoramic scan receiver, at a range that varied according to the plane's altitude and the angle formed by the flight, path, radar site, and tracking beam. The crews discovered that their equipment functioned best when flying at medium altitude, following the beam directly toward the transmitter. On one such occasion, the panoramic scan receiver picked up the tracking signal 107 nautical miles away. The poorest results occurred at low altitudes on a course generally parallel to the radar beam, as one of the F-100Fs was doing when it passed some 13 nautical miles from the set before detecting its transmission. [8]

The Eglin tests demonstrated that Wild Weasel I could detect Fan Song signals beyond the 17-nautical-mile effective range of the Guideline missile, but detection was only the beginning. Next the crew had to locate the radar and its missile launchers. The panoramic scan receiver provided an initial azimuth which the aircraft followed until the signal was strong enough for the shorter range vector homing and warning set.

For a time, the electronic warfare officer could compare the data on the two scopes to ensure that his plane was on the proper heading. As the Wild Weasel closed with the transmitter and came within SAM range, however, he had to rely upon the vector homing equipment, with its launch warning feature, which gave a general azimuth and very rough approximation of range to the antenna. As a result, the pilot then had to search visually for the SAM site before he could attack. [9]

On November 1965, Maj. Gary A. Willard, Jr., a veteran of the Eglin tests, arrived at Korat Air Base, Thailand, with a Wild Weasel Task Force consisting of the four modified F-100F's and their crews. His organization came under the operational control of the 2d Air Division's 6234th Tactical Fighter Squadron, whose F-105D's would join the Wild Weasels on Iron Hand missions, as the hunter-killer attacks against North Vietnamese SAM sites already were known. The task force was to submit reports of these strikes to the Tactical Air Warfare Center at Eglin. [10]

Bad weather delayed the combat debut of the task force. Seven missions were cancelled before the skies cleared enough to permit Wild Weasel I to challenge the defenses of North Vietnam. Finally, on 19 December, two F-100Fs, piloted by Major Willard and Capt. Leslie J. Lindemuth, led flights of F-105Ds into the North, but neither of the Wild Weasel electronic warfare officers, Capt. Truman "Walt" Lifsey and Capt Robert D. Trier, detected Fan Song signals. [11]

F-100F Wild Weasel I aircraft refueling from a Boeing KC-135A Stratotanker over Southeast Asia in 1965. This aircraft was lost in a training accident on 13 March 1966 due to an engine failure. Source: U.S. Air Force

On the following day, 20 December, the task force suffered its first combat loss. Antiaircraft fire downed a Wild Weasel as it was leading an unsuccessful attack on a SAM battery it had located about 5 nautical miles southeast of Kep airfield. An F-105D pilot reported seeing one parachute open before the plane knifed into low-hanging clouds. The survivor was Capt John J. Pitchford, the Wild Weasel pilot, who remained a prisoner of war until his repatriation in February 1973. His electronic warfare officer, Captain Trier, died in the crash. [12](Defense Lion Publication notes that Colonel Pitchford passed away in December 2009. Some reports state that Captain Trier survived the crash but was killed by North Vietnamese troops while resisting capture.)

Captain John Pitchford (left) and Captain Robert Trier (right), the crew of the first Wild Weasel to be lost in action. Source: U.S. Air Force.

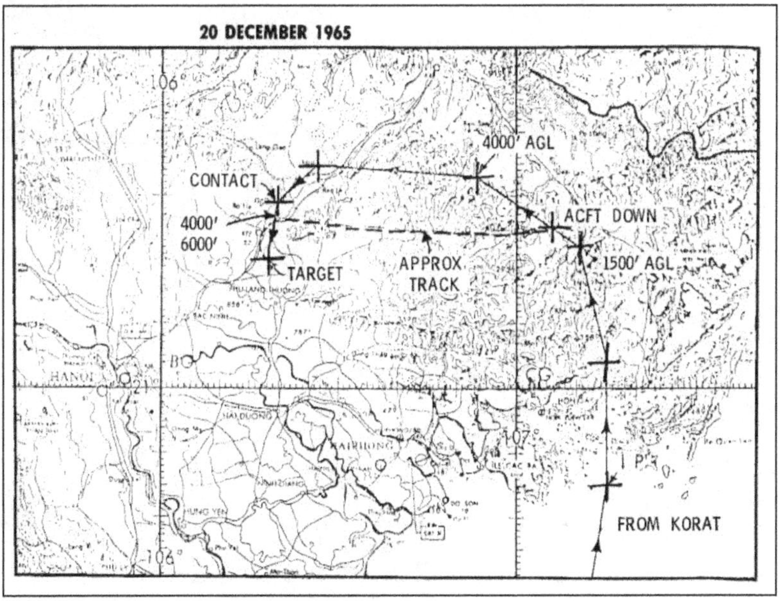

Track chart of Wild Weasel loss on December 20. Source: U.S. Air Force.

Major Willard's men avenged this loss just two days after Pitchford and Trier went down, when one of the Wild Weasels pinpointed a missile site and set it up for the kill. On 22 December, Capt. John E. Donovan, an electronic warfare officer, picked up a distant Fan Song signal as his Thailand-based plane crossed the North Vietnamese border. Turning toward the radiation source, the pilot, Capt. Allen T. Lamb, nosed his F-100F downward until the signal disappeared, climbed until the strobe reappeared on Donovan's scope, and continued porpoising in this fashion until he reached the mountain chain that formed the south-western flank of the Red River valley .

"In the mountains," Lamb reported, "there were a series of four or five valleys that were generally perpendicular to the direction we wanted to go, so I would pull up over a ridge, roll the wings level, and as soon as the EWO called a bearing to the signal I would roll on over using a half barrel roll descending into the next valley, turning sometimes up the valley and sometimes down the valley. " [13]

The Wild Weasel swung around a final hill and climbed' to an altitude of 4500 feet as the pilot began searching for the SAM site. He saw the radar van "sitting in about the center of what I had thought was a village" and then spotted three Guidelines but "only the front part -- the long white tips -- because the missiles appeared to be partially covered by a semi-circular thatched hut."

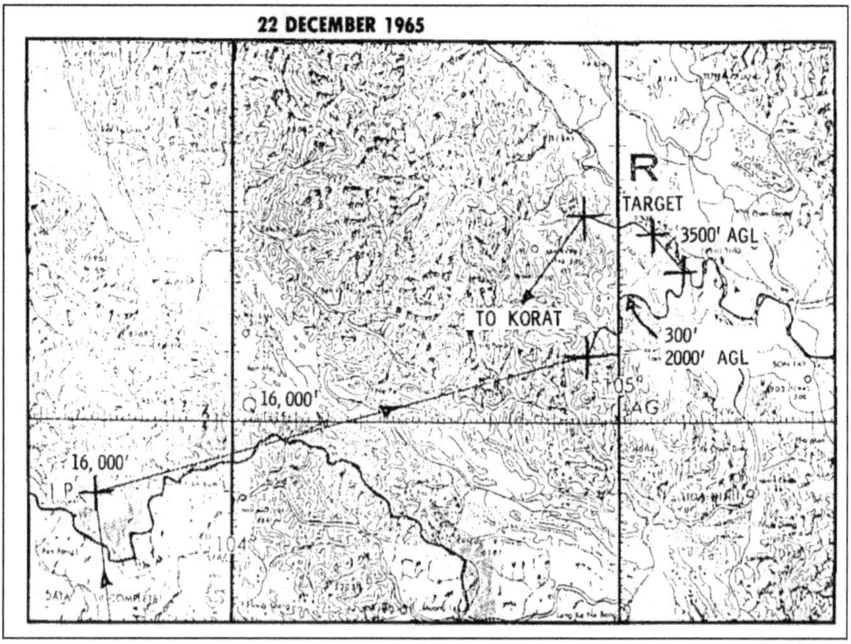

Retribution – the destruction of a SA-2 site on 22 December. Source: U.S. Air Force.

Lamb radioed the F-105Ds, marked the target with rockets, and watched as the fighter-bombers made their passes. Explosions rocked the site, and smoke and dust rose 300 to 400 feet into the air. At this point, Donovan detected a Fan Song signal originating nearby and already in the high pulse repetition frequency. The Iron Hand formation immediately headed back to Thailand, remaining at low altitude until it reached the mountains, which screened the planes from enemy radar. Although the modified F-100Fs contributed to the destruction of just this one SAM site during their combat evaluation, the successful attack demonstrated the potential value of Wild Weasel. [14]

Iron Hand flights used three basic formations during the Southeast Asia tests. One formation consisted of a Wild Weasel hunter and three killer F-105Ds. In both the others, the F-100F located targets for four Thunderchiefs.

If three F-105D's accompanied the F-100F, one of the Thunderchief pilots served as wingman for the Wild Weasel first positioning himself 200 to 1000 feet to the right rear of the formation leader, crossing the leader's wake, to assume position the same distance to the left rear, then recrossing in a mirrored letter S flight path. The other two F-105Ds formed a separate element 2000 to 4000 feet to the left rear of the first Thunderchief, then veered back and forth behind the formation leader, keeping on the opposite flank from his wingman. The Thunderchiefs adopted this weaving maneuver in order to maintain formation behind the slower moving F-l00F.

If four F-105Ds were serving as killers, they had to weave because of the F-100F's slower speed, but they could do so either as individual aircraft, separated from one another by 2000 to 3000 feet, or in pairs, with 2000 to 4000 feet separating the two plane elements. [15]

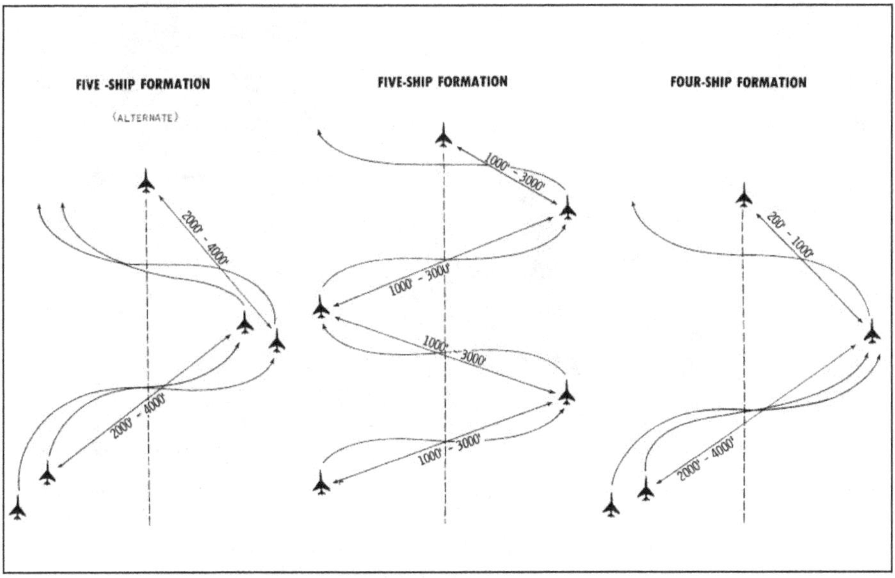

After the evaluation ended on 26 January 1966, the F-100Fs continued to lead Iron Hand flights against SAM sites, but by the end of March they no longer tried to penetrate heavily defended areas in the North. The loss of a second plane to antiaircraft fire apparently had convinced the Second Air Division (which became the Seventh Air Force on 1 April) that the F-100F Super Sabres were too old and too slow to survive in the hostile skies above the Red River delta. Beginning in

May, an improved Wild Weasel arrived in South-east Asia, relieving the converted Super Sabres of their day- time role, though they continued for several months to fly single-plane night missions from Korat. [16]

Meanwhile, back in October 1965, while Wild Weasel I was undergoing evaluation at Eglin, the Air Proving Ground Center and Tactical Air Warfare Center began collaborating on Project Wild Weasel IA, in order to "determine the capabilities of an F-105D aircraft employing radar homing and warning equipment similar to that employed in the Wild Weasel I aircraft. " Should the experiment prove successful, all members of the Iron Hand team would be flying the same airplane, and the killers would no longer have to weave back and forth to avoid outrunning the hunter. Unfortunately, the test merely reinforced a lesson already learned -- simultaneously flying the airplane, operating the electronic gear, and searching visually for SAM sites were too much for one man. [17]

This picture shows just how hard a SA-2 site was to spot visually. Originally, the Russians used a standard star-shaped design that was much easier to see but the Vietnamese quickly discarded it in favor of the irregular layout seen here. Source: U.S. Air Force.

Since the F-105D was not suited to the Wild Weasel role, the two-place F-105F became the logical replacement for the aging F-100F. A modified F-105F, called Wild Weasel II, appeared with new homing and warning equipment mounted in pods on the plane's wing tips rather than stowed within the fuselage. Because the heavy pods caused dangerous vibration, however, the wings of Wild Weasel II had to be strengthened and its indicated air speed kept below 300 knots. This performance restriction, plus the cost of renovation, eliminated the plane from serious consideration as a replacement for Wild Weasel I. [18]

Wild Weasel III, however, a two-place F-105F with essentially the same internally housed homing and warning equipment as Major Willard's F-100F's, passed its test at Eglin, reached Thailand in May 1966, and entered combat the following month. The first contingent of five planes joined the 388th Tactical Fighter Wing at Korat. Six others reached Takhli late in June and by 4 July had entered combat in support of the 355th Tactical Fighter Wing. [19]

Of the 143 F-105F trainers built, 86 were converted into Wild Weasels like the one pictured here. Because of the high losses attributed to such a dangerous mission, though, there were typically fewer than a dozen aircraft available for missions at any one time. Source: U.S. Air Force

The arrival of these F-105Fs, which had the same basic performance as the D model, marked the end of those Iron Hand formations designed to compensate for the difference in speed between the Thunderchief and the F-100F. Hunter-killer flights came to consist of two two-plane elements, at least one of which was led by a Wild Weasel. They generally approached the target in the same loose fingertip formation as a flight of strike aircraft, then separated to cover a broader area. [20]

Plans also called for the deployment to Southeast Asia of Wild Weasel IV. a modified F-4C. This project fell behind schedule because of the difficulty in finding space inside the fuselage for the necessary homing and warning gear. Consequently none of these planes saw action during Rolling Thunder. [21]

Wild Weasel IV was on hand in time for the Linebacker operations of 1972, however, as was the F-105G, which was a modified Wild Weasel III. The G model featured an improved radar warning receiver and a jamming transmitter mounted in a blister beneath the fuselage. Work on the first of F-105Gs began late in 1969. [22]

Radar Suppression

During the operational testing of Wild Weasel I in Southeast Asia, Major Willard's crews first homed on enemy radar signals, then tried to pinpoint the SAM site, using rockets to mark it for the kill. As the loss of two F-100F's attested, finding a camouflaged missile battery was dangerous as well as difficult, for the Wild Weasel crews had to brave antiaircraft fire, fighter attack, and the threat of SAMs throughout the painstaking search. The U.S. Navy, however, had a missile that promised to simplify the hunt for SAM units.

The AGM-45. Source: U.S. Air Force

This weapon was the AGM-45 Shrike, a solid-propellant missile weighing 180 kilograms (400 pounds) that could home on a radar transmitter from a distance of more than 17 nautical miles, the maximum effective range of the Guideline missile. The weapon 1 s sensor head reacted to signals over a specified frequency band and could home on any radar that functioned [9] within this band. A proximity fuse detonated the 63-kilogram (140-pound) warhead. The pilot launching a Shrike had to fly directly toward the radiation source and loft the missile so that its ballistic track would bring it close enough to the target for the homing system to take over. Some 10 seconds after launch, the rocket motor burned out and guidance ceased, but if released precisely the weapon could strike within 20 feet of the radar antenna against which it was directed. [23]

The Wild Weasel I detachment was the first Air Force unit to launch a Shrike in combat. This initial target, attacked on 18 April 1966, was a Fire Can radar located 5 nautical miles northwest of Dong Hoi in the panhandle of North Vietnam. The pilots of the three F-105Ds that formed the killer component of a four-plane Iron Hand team tried to follow the missile in order to bomb the transmitter and gun positions, but the Shrike vanished in the haze. The electronic warfare officer in the back seat of the F-100F saw the radar signal disappear from his scope. Since the transmitter remained mute for the remainder of the mission, he concluded that the Shrike might have destroyed or badly damaged it. [24]

Despite the introduction of Shrike casualties remained high among Wild Weasel crews, whose tactics left them vulnerable to MiGs and antiaircraft fire. Iron Hand teams preceded the main strike force by 5 minutes, which deprived them of protection from the F-4s that defended the main formation from fighter attack. In addition, while the strike aircraft thundered toward the target, dropped their bombs, and withdrew, the Wild Weasels kept searching for SAM sites, sometimes remaining for as long as 35 minutes over an area bristling with antiaircraft guns. [25]

F-105F Wild Weasel attempting to dodge a SA-2. There is no information available on whether this particular crew survived the encounter. Source: U.S. Air Force.

So heavy were the losses that by mid-August 1966 only 4 Wild Weasels remained flyable of the 11 converted F-105Fs dispatched to Thailand earlier that summer. Although the 388th Tactical Fighter Wing had lost only one of its five Wild Weasels, the enemy had shot down four of the six planes assigned to the 355th Tactical Fighter Wing and damaged the other two beyond repair. In October. six replacements arrived and were divided between the wings so that each had five aircraft. Almost immediately, the North Vietnamese downed an F-105F from the 388th Tactical Fighter Wing, flown by the Wild Weasel detachment commander. For the remainder of Rolling Thunder the number of Wild Weasels serving with each wing varied from as few as 4 to as many as 12. [26]

Despite the losses suffered by the units that launched it. Shrike did not destroy many radars. In fact, other types of ordnance carried by the Wild Weasels proved more deadly, once the crews had located the target. From 18 April through 15 July 1966, F-100Fs and F-105Fs launched 107 of the missiles but scored only 1 confirmed and 38 probably hits. Various factors contributed to this unimpressive record, among them the small explosive charge which required that Shrike score a direct hit to inflict mortal damage. The principal reason for the scarcity of hits, however, was the fashion in which the missile had to be launched.

Because the electronic gear on board the Wild Weasel could not determine the precise range to target, the crew had to use some other method of getting the Shrike within homing distance of the enemy radar. A typical launch maneuver consisted of diving toward the transmitter until the missile's radiation seeker had locked onto the proper azimuth and elevation, then pulling up and lofting the Shrike toward its target. A reference table told the crew what loft angle to use for a given speed, dive angle, and altitude. This aiming system had two built-in weaknesses. First, the Wild Weasel was hurtling through the air as the crew made these calculations and was closing with the radar transmitter. Second, the table being consulted was based, for ease of computation, on the assumption that the target lay at sea level; therefore, if the radar was transmitting from a hilltop or ridge line, the loft angle would be incorrect. [27]

In using the launch table, the crew flew directly toward, the intended target, thereby alerting the radar operators to the imminent danger. Fan Song crews soon realized that they could frustrate the Shrike by shutting down and depriving it of a radiation source upon which to home. In this fashion, the enemy further reduced the number of Shrike hits, but in consequence also diminished the accuracy of his own SAMs. Since the radars could not cease transmitting whenever an aircraft dived toward them, North Vietnamese operators had to devise a tactical compromise that enabled them to engage the attacking Americans while offering only a fleeting target to the anti-radiation missile. [28]

The enemy found that an acceptable way of countering Shrike was to reduce transmission time by relying on acquisition radars or ground observers to supply the course and speed approaching American formations. Instead of transmitting for 10 or 12 minutes, Fire Can and Fan Song could remain on the air for 3 minutes or less. The North Vietnamese also learned to recognize hunter-killer teams -- usually four planes -- in contrast to the larger strike formations, and to avoid using gun-laying or missile control sets while Iron Hand was nearby. Radar operators shut down as soon as they detected a Shrike launch, which could be seen on the scope or reported by observers posted in the vicinity, thus depriving the homing device of a target. [29]

As these enemy tactics evolved, Wild Weasel crews found they could no longer cruise about and locate radars with their own detection gear. They had to rely upon intelligence reports in positioning their aircraft to engage suspected transmitters. Also, because the radar signals were so brief, the pilot and his electronic war fare officer sometimes could not use the standard table to calculate a loft angle and had to estimate how sharply to pull up. If the enemy had already launched a SAM, the Wild Weasel crew could only make "a best guess at range" and fire a Shrike "in the hope that the firing would cause the operators to shut down the radar causing the SAM to go ballistic and miss its target. " [30]

Despite the limitations imposed on Shrike by the absence of Wild Weasel ranging equipment, crews did make effective use of the anti-radiation missile. Since the enemy radar operators would reduce their transmissions in order to minimize the danger from Shrike, Iron Hand flights used the threat of this missile to deter the North Vietnamese from tracking accurately the strike formations. To make this menace credible, the Wild Weasels would feint toward suspected radars, or sometimes actually bomb them, and promptly attack any Fan Song or Fire Can bold enough to begin transmitting. Occasionally, the Iron Hand teams disguised themselves by flying close to the strike force, hoping to be mistaken for part of it, in order to lull the enemy into going on the air long enough to provide a target for the Shrike. [31]

Between the summer of 1966 and the spring of 1967, the Wild Weasel mission underwent subtle change. At first, emphasis was upon hunter-killer attacks to search out and destroy missile positions, but radar suppression gradually took precedence over destruction, although Wild Weasel crews continued to attack with bombs, rockets, or 20-mm cannon whenever they sighted a SAM battery. Aircraft diverted because of bad weather from Fan Song suppression flights over the Red River delta carried out a secondary hunter-killer against SAM sites in North Vietnam's panhandle. [32]

Among the more celebrated Wild Weasel missions was the one led by Capt. Merlyn H. Dethlefsen against the radar-controlled defenses of the Thai Nguyen steel works, some 40 nautical miles north of Hanoi. When he took off on 10 March 1967, he carried a pair of Shrikes, plus bombs, and the 20-mm cannon built into his F-105F. He had "questioned the amount of damage that an AGM-45 [Shrike] would do by itself" and hoped that he could "actually accomplish something" with bombs and gunfire.

Flying well ahead of the strike force, Dethlefsen encountered antiaircraft fire so dense that he lost sight of the other Iron Hand aircraft among the clouds of smoke from bursting shells. "The sky was just black," he said later. "It was just horrible . . . . . You know you're not bullet proof . . .when explosions are rocking your wings and you can hear metal hitting metal." [33]

Capts. Merlyn Dethlefsen (l) and Mike Gilroy. Source: (U.S. Air Force

Capt. Kevin Gilroy, the electronic warfare officer, located a Fan Song transmitter, but just as Dethlefsen launched, a Shrike toward it, a pair of MiG-21s jumped the Wild Weasel from behind. One of the interceptors launched a heat-seeking missile, forcing Dethlefsen to dive through a carpet of flak to avoid it. Despite the danger from fighters and antiaircraft guns, he stayed in the vicinity of Thai Nguyen as long as fuel remained in an attempt to maintain radar suppression. When Gilroy detected another Fan Song, Dethlefsen used his second Shrike to silence it. He then spotted a radar van parked at a third SA-2 site and attacked with bombs and cannon fire. [34]

Capts. Dethlefsen and Gilroy in their F-105F in April 1967. Against the odds, both Dethlefsen and Gilroy completed 100 missions over North Vietnam. Source: U.S. Air Force

Although Gilroy cannot recall the incident Dethlefsen has said that when the two men landed at Takhli, the fighter pilots based there were "having a big celebration because they'd shot some MiGs and at any rate we were largely ignored. [sic] The accomplishments of the Wild Weasel crew did not remain unnoticed, however. Early in 1968, Dethlefsen received the Medal of Honor for his part in the attack on Thai Nguyen, and Gilroy got the Air Force Cross.[35]

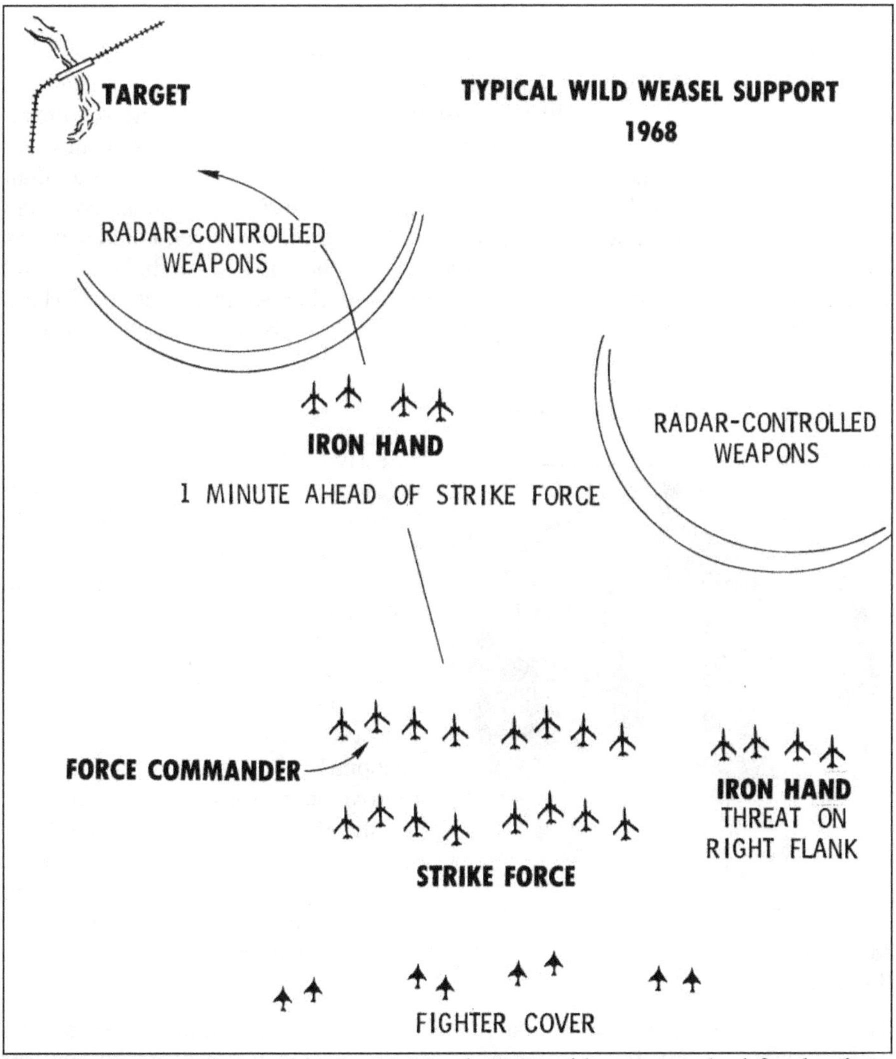

The Iron Hand tactics used in the 10 March 1967 raid were standard for the time. Dethlefsen's F-105F was 1 of 2 which, with a pair of F-105Ds, preceded the strike force by about 5 to 7 minutes (30 to 45 nautical miles) in order to check the weather and suppress the Fan Song or Fire Can radars that helped defend the steel mill. The Iron Hand team was in a vulnerable position. Separated from the larger formation and its F-4 fighter escort, Dethlefsen's group risked attack by hostile interceptors, and both antiaircraft barrages and optically aimed fire awaited the pilot who ventured within range. Unless the radar suppression team remained in the area, as Dethlefsen did, enemy operators could wait until Iron Hand had passed, then resume transmitting in time to direct missiles and gunfire at the fighter-bombers.

By mid-1968, however, Seventh Air Force had revised these tactics. The main strike force now followed just 1 minute behind the Iron Hand flight that led the attack, and a second Iron Hand team usually accompanied the fighter-bombers, flying near the rear of the formation or on the flank where radar-controlled weapons posed the greater threat. The two flights could cover a larger area than one, suppressing several radars simultaneously and ensuring continuous coverage throughout approach, attack, and withdrawal. One team, moreover, could serve as a decoy, tricking the enemy into using radar after it had passed, so that the second would have a target for its Shrikes. Despite these changes in tactics, Iron Hand crews still had to run a gantlet of antiaircraft fire during bombing or strafing runs.[36]

An Improvement over Shrike

As early as the fall of 1966, according to Col. Arthur D. Thomas of the 460th Tactical Reconnaissance Wing, Wild Weasel was giving American Strike forces "some freedom of action in the SAM defended areas," even though the enemy's skilled use of camouflage and Wild Weasel countermeasures" would require the "immediate development of improved Weasel equipment and weapons."[37] Among the major improvements was a new anti-radiation missile, the AGM-78, also called the Standard ARM. A product of Air Force-Navy collaboration, the new weapon made its combat debut late in March 1968, too late to have much effect on Rolling Thunder, which ended within 8 months.

The Standard ARM had a longer range than Shrike, greater destructive power, and better homing ability. In theory, a pilot could engage a target 60 to 70 nautical miles distant, but he would need luck to obtain a fix on a radar that far away. Also, to attain this range, he would have to release the missile from an altitude of 40,000 feet, while in actual practice the Wild Weasels usually launched the AGM-78 from 10,000 to 24, 000 feet. The Standard ARM boasted a warhead weighing 99 kilograms (219 pounds), almost half again the weight of Shrike's explosive charge. The improved homing system, which enabled the pilot to avoid flying directly at the target, contained a memory circuit that kept the missile on course even though the enemy radar had ceased transmitting a few seconds before the time of intended impact. On the debit side, the 610-kilogram (1,350-pound) AGM-78 was more than twice as heavy as Shrike and, in the limited numbers manufactured by the spring of 1968, was roughly 10 times as expensive, with each of the new missiles costing $200,000. [38]

The AGM-78 Standard ARM in front of an F-105G. The missile under the Thud's wing on the right is an AGM-45 Shrike. Although a size comparison is distorted by perspective, the size of the AGM-78 is obvious. The picture also shows the ECM pod scabbed on to the fuselage of the F-105 where the bomb bay used to be. Source: U.S. Air Force.

From 1 April, when President Johnson halted the air war north of the 20th parallel, until Rolling Thunder ended completely on 1 November, Wild Weasels used both Shrikes and Standard ARM missiles in missions against southern North Vietnam. For sorties near the demilitarized zone or along the seacoast, Shrikes proved preferable to Standard ARMs because the longer range missile might endanger friendly forces. During this same period, Wild Weasel crews revived the practice of trolling, flying near suspected SAM sites to entice Fan Song operators into transmitting. When a radar came on the air, a Shrike or AGM-78 promptly homed on it. Should the SAM battery succeed in launching a missile, the Wild Weasels could dodge it, look for a cloud of dust raised when the first stage ignited and bomb and strafe the site. [39]

The Pay-off. A Vietnamese SA-2 site vanishes under the smoke of multiple bomb and rocket hits. Source: U.S. Air Force.

Wild Weasel and Radar Bombing: Ryan's Raiders and BASS

Wild Weasel crews and aircraft took part in two radar bombing programs, one successful, the other a failure. The successful venture originated with Gen. John D. Ryan, USAF, who assumed command of Pacific Air Forces in February 1967 and almost immediately began searching for some means of continuing Rolling Thunder attacks during darkness and bad weather. His initiative resulted in the modification of several F-105Fs to fly either night bombardment or conventional Wild Weasel missions. Because of the general's role in setting up the project, the men who carried out the night strikes called themselves "Ryan's Raiders."[40]

On the night of 6 April, Ryan's Raiders struck for the first time, bombing a ferry and rail yard deep inside North Vietnam. Raids continued throughout the remainder of Rolling Thunder, but the night operation remained essentially a form of harassment. Too few planes were available for a systematic offensive, and those actually used could not attain pinpoint accuracy, since they relied upon a radar bombing system originally designed for nuclear weapons rather than 340-kilogram (750-pound) bombs.[41]

When Ryan's Raiders made their first night strike, eight F-105Fs modified at Yokota Air Base, Japan, was able to perform either night attack or normal Wild Weasel operations. Two of these planes were lost in May, but the 388th Tactical Fighter Wing managed to replace them with a pair of F-105Fs modified at Korat. Seven dual-purpose Thunderchiefs survived beyond the year's end, to be joined in February by the first of six F-105Fs equipped exclusively for nighttime radar bombing. The dual-purpose aircraft reverted to Wild Weasel duties, though they remained available to Ryan's Raiders as replacements for planes being repaired. [42]

At the outset, Ryan's Raiders 1 air crews consisted of two qualified pilots. Since both officers had flown the single-seat Thunderchief, they were familiar with the radar bombing equipment common to the D and F models. The detachment had been in action less than a month, however, when Lt. Gen. Joseph H. Moore, Vice Commander of Pacific Air Forces, decided that two pilots were a luxury and proposed training Wild Weasel crews, made up of pilot and electronic warfare officers, for night attack. His plan benefited from a pool of the SAC-trained electronic warfare officers, actually countermeasures specialists, who mastered the F-105F radar navigation and bombardment equipment within a few weeks. Beginning in July 1967, four Wild Weasel crews alternated between daylight radar suppression and night bombing. Following the appearance early in 1968 of the F-105Fs equipped solely for radar bombardment, the Wild Weasel crews were relieved of night bombing duties, but they had demonstrated such skill in night operations that Pacific Air Forces did not revive the original requirement for two-pilot crews. [43]

In addition to providing the volunteer night bombing crews, Wild Weasel units sometimes performed nighttime radar suppression as Ryan's Raiders flew low-altitude, night-time penetrations deep into North Vietnam. Fitted with an auxiliary fuel tank and just one Shrike missile, a Wild Weasel circled at high altitude beyond range of SAM batteries, while an F-105F from Ryan's Raiders challenged the radar-controlled defenses as it hurtled through the darkness. At first, the lone Wild Weasel may have diverted attention from the attacking aircraft, and even though radar operators soon realized that the appearance of one plane high in the sky heralded the approach of a second at low altitude, they seemed reluctant to transmit at the risk of attracting a Shrike. [44]

The radar bombing scheme that failed bore the nickname BASS, an acronym for Bistatic Aided Strike System, and employed a specially equipped Lockheed EC-121K, called Rivet Top, to direct modified Wild Weasel against North Vietnamese radar. Inside Rivet Top's cramped fuselage, a controller first pinpointed an enemy radar, then watched for the signal from a transponder mounted on the Wild Weasel which had been triggered by the hostile transmission. The controller tried to coach the Wild Weasel on the transmitter, telling the pilot what headings to follow and when to attack. Luring 10 combat tests conducted in September 1968, however, the Wild Weasel's transponder failed to register on the controller's scope, and he could not guide the plane to its target. [45]

An Appraisal of Wild Weasel

Assessing the effectiveness of radar suppression was a problem from the outset. The first Shrike launched over North Vietnam had vanished into a ground haze, leaving no clue to its effectiveness except the fact that the radar set ceased transmitting. Although individual crewmen felt otherwise, intelligence analysts could confirm only one hit by the first 107 Shrikes launched against North Vietnamese radars. Yet, several of these transmitters were destroyed or damaged by Wild Weasel crews who spotted them and attacked with bombs, rockets, or gunfire.

Do NOT annoy the Bear in his den. The back seat of an F-105G. Source U.S. Air Force.

As enemy radar operators became more skillful in limiting transmission time, the Wild Weasels found it increasingly difficult to pinpoint targets electronically and sometimes launched against brief bursts of radiation from sources invisible to them. Once the crews had seen the target, however, these radar techniques were useless against the more conventional munitions. [46]

The fact that a radar attacked by a Shrike or Standard ARM might then be bombed by the very aircraft that had launched the missile complicated the problem of judging the effectiveness of the anti-radiation weapons, for no one could declare with certainty what type of ordnance had destroyed a target. Also, because of the speed with which the enemy could replace a radar knocked out by aerial attack, intelligence analysts sometimes found it difficult to determine whether the Wild Weasel had actually scored a hit.

The Air Force Security Service, the agency responsible for determining the impact of electronic countermeasures, had access to radar intercepts that gave insight into Wild Weasel effectiveness. For example, the 35 Shrikes launched in southern North Vietnam from April through June 1968 caused no discernible damage but they did force the enemy to transmit briefly and at irregular intervals, undercutting his ability to burn through American jamming. The 45 Shrikes launched during the last 90 days of Rolling Thunder similarly disrupted radar coverage, even though only three of the missiles actually damaged North Vietnamese radar vans. Wild Weasel did therefore suppress enemy radars despite the small number of hits sea red with anti-radiation missiles; it sometimes accomplished its purpose by showing its fangs rather than drawing blood with them. [47]

IV. "THE MOST SIGNIFICANT DEVELOPMENT"

Col. Robin Olds, commander of the Thailand-based 8th Tactical Fighter Wing from September 1966 to December 1967, declared that "the most significant development in the air war over North Vietnam during my tour was the introduction of the ECM [electronic countermeasures] pod and with it a return to mass formation tactics reminiscent of fighter operations in World War II and Korea." [1]

Brig. Gen. William S. Chairsell, commander of the 388th Tactical Fighter Wing from August 1966 to August 1967, made a similar evaluation. "Prior to the ECM pod, " he wrote, "our aircraft were required to ingress and egress to and from the target using terrain masking for protection and employ the 'pop-up' maneuver over the target," tactics that brought the planes "well within range of the majority of AW [automatic weapons] and AAA [antiaircraft artillery]" and made them "extremely vulnerable to SA-2 firings and AAA at the peak of their pop-up." Thanks to the self-protection pod, General Chairsell continued, "our aircraft could now roll into the target from medium altitude -- 12,000 to 15,000 feet," a change that reduced losses and improved bombing accuracy. [2]

Overcoming Early Disappointment

The first pod used in Southeast Asia, the QRC-160-1, gave no indication that its successors would earn such tributes. Although, the Air Force had been working on the idea since 1959, the early device was a failure. The internal construction could not withstand the stresses created by the high-speeds of the RF-101s that carried it, and the external shape generated aerodynamic forces that twisted the plane's wing tips. Consequently, after a brief trial in the spring of 1965, the pod was withdrawn from service. [3]

QRC-160-1: Source: U.S. Air Force

Not all of the early self-protection equipment proved disappointing, however, Navy airmen jammed Fan Song using the ALQ-51, a deception device which transmitted a false position when triggered by radar waves. Instead of being enclosed in a detachable pod like the QRC-160-1, the ALQ-51 was built into the underside of the fuselage of Grumman A-6s and late-model

McDonnell Douglas A-4s. On 16 September 1965, an A-4E pilot reported success in jamming a Fan Song beam that had been tracking him. Four days later, a flight of A-6's used their deception jammers in escaping a half-dozen SAMs launched against them. [4]

The Air Force U-2 also carried a deception jammer, which was successful against Fan Song during flights over the North. U-2s were few in number, however, and suited only for high altitude reconnaissance, so the failure of the QRC-160-1 placed the burden for Air Force countermeasures on the EB-66. But the EB-66 unfortunately was incapable of accompanying strike forces into heavily defended areas and therefore could not effectively assume countermeasures responsibilities. In the words of Col. Arthur R. Thomas, what the Air Force needed was "a penetration aid that would defeat SAM and radar systems throughout the mission." As a result, correcting the defects in the QRC-160-1 received a high priority. [5]

ALQ-71 (previously, the QRC-160-A-1) pod on an F-105 wing pylon. Source: U.S. Air Force.

In January 1966, the Air Proving Ground Center at Eglin AFB began testing the QRC-160A-1. This modified pod, though structurally stronger and somewhat heavier than the original, had undergone no radical change. Like its predecessor, it was a self-contained, barrage noise jammer with four 75-watt magnetron transmitters (soon replaced by 100-watt models). These components fitted inside an aerodynamic pod measuring 230 centimeters (7.5 feet) in length by 25 centimeters (10 inches) in diameter. Total weight was 90 kilograms (200 pounds). Except for a 28 volt status light in the cockpit, all electrical power came from an integral ram-air turbine.

Like the older model, the improved self-protection pod was simple to operate. Prior to takeoff, ground crewmen adjusted the controls on the outside of the pod to establish the center frequency and band width that would jam particular kinds of radar. The pilot needed only to turn the transmitters on and off. [6]

Preliminary evaluations indicated as early as the end of January 1966, that the QRC-160A-1 would prove rugged enough for combat and was effective against Fan Song or Fire Can. By mid-year, the test results convinced Gen. William W. Momyer, who had recently assumed command of the Seventh Air Force, that the new self-protection pod offered a means of reducing losses over North Vietnam. He therefore requested a combat evaluation, and 25 of the pods, with technicians to maintain them, were sent to Thailand. In September, the test team and its equipment reached the 355th Tactical Fighter Wing at Takhli which immediately launched Project Vampyrus to determine the value of the pods against radar-controlled weapons. [7]

Project Vampyrus [8]

From 26 September through 8 October 1966, F-105Ds of the 355th Tactical Fighter Wing, commanded by Col. Robert R. Scott, flew 19 four-plane missions, sometimes as many as three a day, to test the pods against targets defended by SAMs and radar controlled antiaircraft guns. Maj. Douglas D. Brenner, of the wing's 333d Tactical Fighter Squadron, served as operations officer for Vampyrus, Capt. Karl G. Berroth supervised maintenance, and troubleshooter for electronic problems was Capt. David S. Zook, who had served as project officer during the Eglin testing.

The project moved forward with EB-66Cs measuring the effectiveness of the pods. The electronic warfare officers on board first verified the number, location, type, and transmission characteristics of radars protecting the targets, then observed how these transmitters reacted to the pod-carrying Thunderchiefs. When necessary, the EB-66C's jammed radars outside the frequency range covered by the self-protection pods. Besides testing the mechanics of the pods, the Vampyrus task force experimented with formations, trying to lay down the densest possible noise barrage with the least sacrifice of maneuverability. At first, the leader flew at altitudes varying from 6,000 to 16,000 feet, with others echeloned downward, each pilot remaining 1,000 feet below and 1,500 feet behind the plane ahead. This formation proved awkward however. The fliers complained that they found it hard to locate navigational checkpoints and bombing targets because they had to look up so frequently to maintain station. To avoid this distraction they decided to form an echelon extending up and back from the leader. And, to improve further the chance of picking up the target, the leader sometimes maintained an altitude of

4,500 feet, the lowest he could fly without running suicidal risks from light antiaircraft guns.

The most spectacular Vampyrus mission occurred on 8 October 1966 when three four-plane flights of F-105Ds made coordinated strikes against the Nguyen Khe oil storage facility south of Hanoi. A pod-carrying Thunderchief of one of these flights -- Taksan flight -- had mechanical trouble and was replaced by a plane that had no jamming equipment. Prior to entering the target area at an altitude of 4,500 feet, this flight divided into two elements, and almost immediately the two pods carried by Taksan 01 ceased functioning. The other plane, Taksan 02, was the last-minute replacement and had no countermeasures protection at all.

Within seconds, the pilot of Taksan 01 found himself in mortal peril. A MiG-21 dived toward him, and as he eluded it, three 85-mm shells burst close enough to punch holes in the skin of his F-105D. He jettisoned his bombs to gain speed and maneuverability, but his engine suddenly quit. No sooner had he restarted the balky turbine than a SAM came streaking toward him. He quickly nosed over in a diving turn that brought him within 2500 feet of the ground. Antiaircraft shells burst around him until he climbed out of range. Taksan 02, also without countermeasures protection, encountered comparably intensive fire from radar controlled antiaircraft guns but luckily escaped damage. The other two planes in the flight, however, Taksan 03 and 04, turned on their self-protection pods and experienced only light to moderate fire.

For the second Vampyrus flight, Drill the approach to the target was more dangerous than the actual bombing. The four planes thundered toward their goal at 2,000 feet, boring through planned barrages from machine guns and light anti-aircraft. Since this fire was not controlled by radar, the pods did not affect its accuracy. Jamming became critical after the F-105Ds climbed to 12,000 feet, just minutes from the target, and exposed themselves to weapons directed by Fan Song and Fire Can. The formation, however, encountered no opposition from SAMs or antiaircraft guns as it dive bombed the oil storage tanks and returned at medium altitude to Thailand.

The four pod-carrying F-105Ds of Steel, the third Vampyrus flight, led by Major Brenner, now arrived on the scene. Staggered between 12,000 and 16,000 feet, they took no evasive action but headed straight for the storage area, dropped their bombs, climbed 4000 feet, and set a course for Takhli. None of the pilots saw either SAMs or bursting shells.

The Vampyrus project proved to the satisfaction of the participants that pod-carrying Thunderchiefs, flying in loose formation at medium altitude, could successfully defy radar-controlled weapons. Pilots flying below 4000 feet continued to face danger from barrage fire by antiaircraft guns and light automatic weapons, as the experience of Drill flight had shown.

The tests also indicated that the QRC-160A-1, soon to be redesignated ALQ-71, was sufficiently reliable for employment in combat operations. Maintenance had been a problem, however, with the ram-air turbine a recurring source of trouble. The next step was to acquire enough of the pods to equip all the fighter- bombers and tactical reconnaissance planes routinely testing the defenses of the North.

Major Advantages and Minor Problems

"Putting these noise jamming pods on each of four aircraft which fly a rather precise, widely spaced formation presents . . . a large, spatially dispersed source of noise and denies range and good direction information to the radars," according to William R. Rambo, director of Stanford University's electronic laboratory. Mr. Rambo cited some impressive statistics concerning pod effectiveness, telling the Air Force Scientific Advisory Board that before Vampyrus the North Vietnamese were averaging one kill for every 35 SAMs launched, but that afterward, the ratio was one plane downed per 60 launchings. [9]

A fighter-bomber pilot confirmed these statistics. "When I got my orders in 1967, 11 said Lt. Col (then Maj.) Robert Belli, USAF, "the sortie rate until you could expect to get shot down was 66 missions for a '105' pilot." But, he discovered, "we had very few shot down the whole time I was there, in a year maybe seven or eight fellows went down." To Belli, it seemed that "electronic warfare just turned the attrition rate upside down, from being pessimistic to optimistic." [10]

Despite its impact upon the loss rate, the self-protection pod did not confer absolute immunity. During level flight, two pods protected an aircraft to within 8-10 nautical miles of the SAM site. At this point, "burn through" occurred, as the Fan Song beam overpowered the jamming signal, enabling the controller to locate a target for his missiles. Also, because of the aircraft's antenna propagation pattern, jamming coverage decreased markedly during maneuvers, especially in steeply banked turns when the strongest signal might not blanket every radar-controlled weapon that could track the target and open fire. As a result, pods might blind one radar, while at the same time leaving holes in the jamming pattern through which another set could lock onto the aircraft and destroy it with SAMs or gunfire. [11]

The ALQ-71 had a few design shortcomings which escaped notice until the pods became standard equipment early in 1967. One such problem was the location of the control box. This was installed to the F-105D pilot's right rear, where it was difficult for him to see the light that indicated failure of the jamming device. The pod therefore might quit functioning without his realizing it. Maintenance men in Thailand resolved the difficulty by moving the control box.

At the same time, they corrected another failing, a matter of mutual interference, by rewiring the box. Before this modification the same jamming signal that disrupted enemy radar also prevented the pilot from using his homing and warning gear. The change in wiring enabled him to interrupt the jamming for a few seconds, long enough for the homing and warning gear to react. [12]

Iron Hand teams had to exercise caution when using both jamming pods, and anti-radiation missiles because the ALQ-71 reduced the accuracy of the Shrike homing device and created clutter on the radar homing and warning scope in the Wild Weasel hunter aircraft. Iron Hand crews therefore tried to avoid turning on the pods until after receiving SAM warning and launching the anti-radiation missile. [13]

By January 1967, the Seventh Air Force had enough ALQ-71s and mounting brackets to begin equipping F-4 Phantoms as well as the F-105Ds. Unlike the Thunderchiefs, which had wiring for two pods, the F-4s could carry only one, and this single jamming device was at best marginally effective. With only half the number of pods carried by Thunderchiefs, a four-plane flight of F-4s sacrificed so much jamming coverage during steeply banked turns that the noise barrage scarcely inconvenienced skilled radarmen. By mid-year, however, electronics technicians had modified the Phantoms to carry two pods and jam as effectively as the F-105s. [14]

The battle between SAMs and countermeasures never ended. These Israeli F-4s seen in 1973 are carrying three ALQ-71 pods each. The antenna set-up has changed. On certain frequencies, each pod would have a long & short antenna, both dedicated to the same frequency. The reason of the two lengths was because of the way the antenna "responded" to the threat. Source: Israeli Defense Forces.

Although the RF-101 Voodoo had introduced the unsuccessful QRC-160-1 pod to Southeast Asia, this reconnaissance plane did not at first carry the improved ALQ-71. Since the Voodoo usually flew alone, refusing to jam unless actually tracked by hostile radar, it seemed ideally suited to a different self-protection jammer, the ALQ-51. The ALQ-51, when triggered automatically by enemy signals, broadcast a false radar return to deceive enemy radar operators. The North Vietnamese, however, blazed away so furiously at the false targets presented them by the ALQ-51 that damage to the aircraft seemed inevitable, and Seventh Air Force had to substitute the ALQ-71. [15]

In November 1967, SAMs scored eight kills within just 4 days, despite the self-protection pod. An investigation team flew to Southeast Asia and discovered that a combination of faulty radar bombing technique and the cunning of North Vietnam's defenders had been responsible for the startling losses. Some Fan Song radars were transmitting on a slightly lower frequency, thus escaping the jamming barrage. Also, Air Force planes tended to bunch up in order to obtain a more compact bomb pattern during missions controlled by ground-based radar, and in doing so they simplified the task of tracking and aiming at the jamming source. Instructions immediately went out to open up formations and to adjust the center frequencies of noise barrages directed against Fan Song. [16]

Tracking the jamming source (also called passive tracking) enabled the enemy to diminish the effect of self-protection pods. Instead of relying on the radar return, which the noise barrage had obliterated, the enemy followed the source of this noise on his radar scope, then launched missiles at the center of the jamming pattern after verifying the range. To ensure the correct range to the target, the North Vietnamese used triangulation, transmitting from widely separated radar sites linked by radio or telephone. The radars operated just long enough to pinpoint the aircraft and a really expert Fan Song operator could avoid using the tracking beam and guidance signal until the SAM had risen from the launcher. Although there was some sacrifice of accuracy, these tactics reduced the impact of the ALQ-71, decreased SAM warning time, and minimized exposure to anti-radiation missiles. [17]

Standardizing Formations

Because the amount of power generated by a self-protection pod was limited, formation flying was essential so that several devices could reinforce one another. Too tight a formation, however, created a compact jamming source that invited fire from an enemy skilled at passive tracking. From the first Vampyrus missions, pilots had experimented with different tactical alignments, but even the most effective formation from the standpoint of electronic coverage had certain tactical disadvantages. As analysts at Seventh Air Force headquarters pointed out, "the pod formation, while optimizing electronic countermeasures, complicates the dive bombing delivery problem." These specialists also reported a "pronounced shallowing effect on dive angles" because only the leader attained "the desired roll-in point."

Pilots realized they could improve accuracy by abandoning the pod formation about a minute before rolling into their dives. When necessary, the pilot farthest to the right shifted to the leader's left in order to place himself on the side opposite the target and clear the line of vision of the other two members of the flight.

Despite such last minute adjustments, dive bombers sometimes had to attack at angles as shallow as 30 degrees instead of the ideal 45 to 60 degrees. [18]

The unavoidable sacrifice of accuracy was acceptable because of the protection afforded by the countermeasures pod. Except when actually dive bombing a target, pilots accepted the discipline of a pod formation. The positioning of the aircraft varied according to unit, however, for the two wings that did most of the Rolling Thunder bombing adopted slightly different formations.

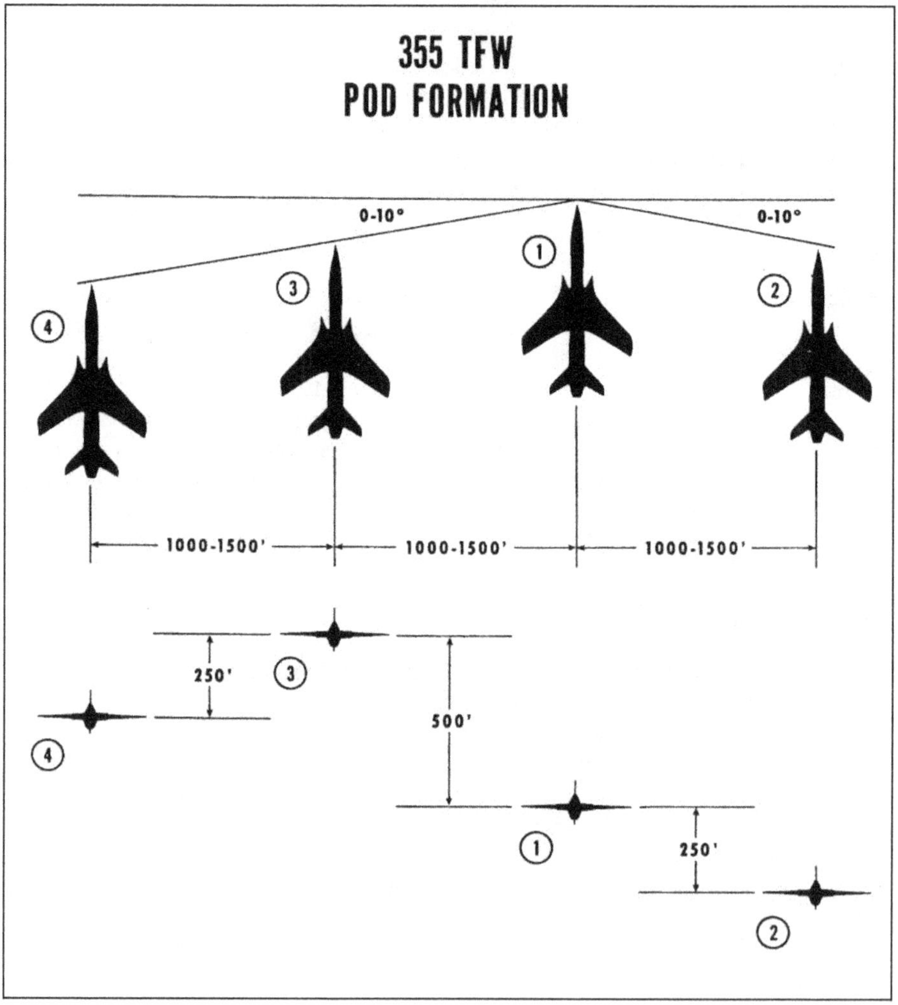

By the summer of 1968, the 355th Tactical Fighter Wing at Takhli was dispatching pod formations of four aircraft, consisting of a pair of two-plane elements, extending 3,000 to 4,000 feet across and 750 feet from top to bottom.

Lowest and farthest to the right was the 02 aircraft, with the formation leader, 01, 250 feet higher and slightly forward of this plane. Echeloned upward from the leader and 10 degrees to his left were planes 03 and 04, with 03 500 feet higher than the leader and 250 feet higher than 04. (When the QRC-160A-1 improved pods first made their combat debut, Vampyrus pilots flew a less precise formation that extended 3,000 feet from top to bottom, with the aircraft echeloned 1000 feet apart.)

The 388th Tactical Fighter Wing at Korat adopted a deeper formation than the 355th, one that extended 3500 feet across and 1500 feet from top to bottom. Again, 02 on the far right was lowest, with the leader 500 feet higher and slightly ahead of him. Numbers 03 and 04 were echeloned at regular intervals, 10 degrees to the leader's left rear, 500 feet separating them vertically.[19]

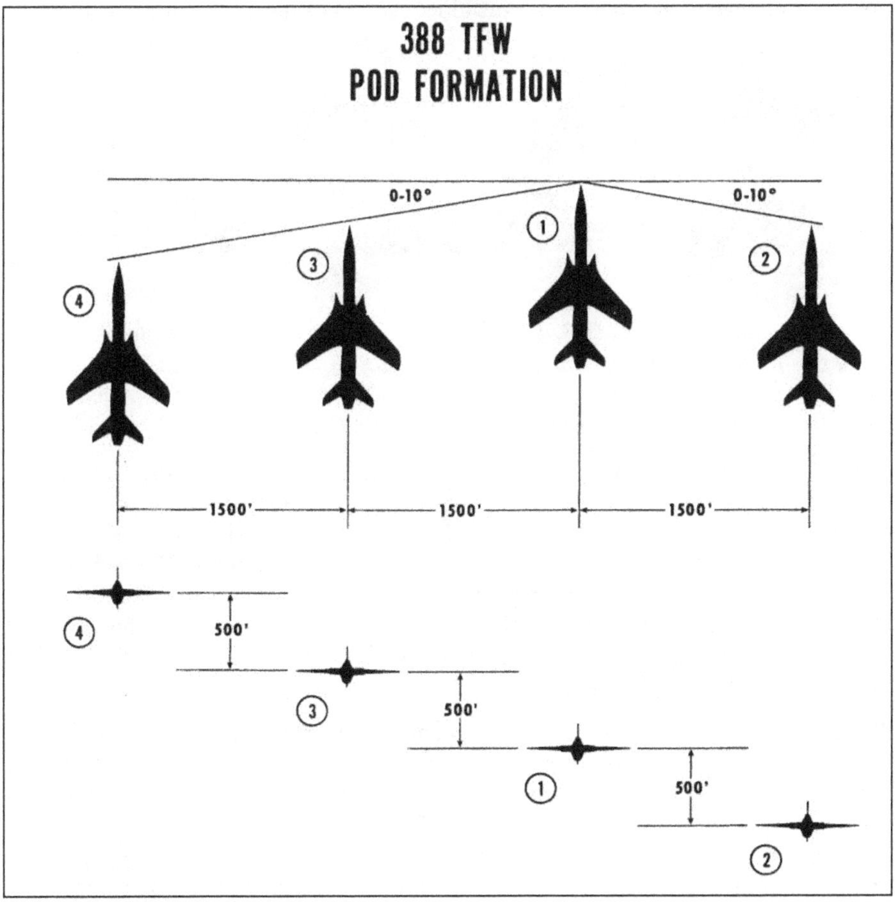

New Equipment

As these formations were evolving, new kinds of self- protection pods arrived in Southeast Asia. The most important of them was the ALQ-87 (originally called QRC-160-8). first used in combat late in 1967 and in general use early the following year. At first, a ram-air turbine supplied current for the four magnetron jammers housed in the ALQ-87, but this power source proved too frail to withstand high-speed maneuvering, so the pod had to be tied into the plane's electrical system. Besides laying down a continuous jamming barrage, the ALQ-87 had a sweep modulator which could introduce random bursts of reinforcing noise in a so-called "pulse power option." The pod, therefore, could simultaneously perform any two of three functions: denying range and azimuth data to Fire Can; depriving Fan Song of range, altitude, and azimuth; and jamming the position beacon installed in the sustainer section of the Guideline missile. [20]

F-111s equipped with ALQ-87s under their fuselages. Close examination shows that this picture has been edited to remove details of the equipment. In addition, the aircraft in the background is carrying something that has been removed by the censor. Source: U.S. Air Force.

Beacon jamming, called down-link jamming, interfered with the signal that enabled the SAM controller to follow the missile on radar in order to correct its trajectory. Both the ALQ-71 and ALQ-87 enjoyed impressive success with this technique. For example, from the inception of beacon jamming in December 1967 until 1 April 1968, when bombing north of the 19th parallel was banned. SAM batteries launched some 495 missiles at Iron Hand Thunderchiefs but downed only three planes and two of the planes had been jamming the tracking beam instead of the down link. [21]

The next piece of self protection equipment to appear in Southeast Asia was the QRC-335, which could perform either deception or noise-barrage jamming. Its designers intended it for activities, such as Iron Hand, where pod formation was not feasible. Consequently Wild Weasel F-105Gs had this transmitter installed beneath the fuselage. The device, however, did not enter service until the April 1968 bombing restrictions were in effect and therefore had only slight impact on Rolling Thunder. [22]

Jamming Enemy Communications

Although the self-protection pods saw extensive service, the Air Force made only limited use of communications jammers, such as the QRC-128 transmitters in the EB-66s. The standard strike formations were designed for pod protection against fire control radar, and planners showed less concern about disrupting communications among ground controllers and MiG pilots. SAMs and antiaircraft guns, after all, posed the greatest threat during Rolling Thunder; also, communication jamming interfered with friendly as well as hostile very-high-frequency (VHF) radio traffic. [23]

Despite the problems, EB-66s, along with Navy and Marine Corps planes carrying AN/ALQ-92 or AN/ALQ-55 transmitters, made occasional attempts to disrupt enemy communications. To avoid drowning out friendly messages, electronic warfare officers waited for a "start jamming" code word, unless a MiG was approaching their aircraft. Another code word, also issued over a prescribed ultra-high-frequency channel signaled 'cease jamming.' [24]

In addition to the EB-66's, eight Takhli-based F-105Fs carried the QRC-128. Too bulky to be placed in a pod, this equipment took almost the entire rear cockpit, replacing the electronic warfare officer, who was not needed because the QRC-128 responded automatically to a predetermined radio signal. The bombing restrictions of 1 April 1968 confined the air war to regions lightly defended by radar-controlled MiGs, so these modified F-105Fs saw little action.

Following the limitation on bombing, two Navy aircraft, an EKA-3B and an EA-6A, responded with their AN/ALQ-55 jammers when a pair of MiGs attempted to intercept a Navy Ling Temco Vought RF-8G reconnaissance craft and its F-8E fighter escort north of Vinh. An Air Force Electronic Warfare Center evaluation indicated that the jamming was effective. One of the MiGs, intent on pursuing the photo plane, apparently failed to receive a radioed warning of the F-8E, which shot him down from behind. On the following day, 10 July 1968, a Navy F-4J destroyed a MiG-21 under similar circumstances. [26]

On 22 September of that year, an Air Force EB-66E joined an EA-6A in jamming very-high-frequency radio communication as two MiGs bore down upon American aircraft conducting strikes in the panhandle of North Vietnam. The guided missile cruiser USS Long Beach requested the jamming, recalled friendly fighter-bombers, and launched a Talos missile that downed one of the interceptors. Analysis indicated that communications jamming might have prevented the victim from receiving a radioed warning of the missile firing. This was the last apparent success for the type of electronic countermeasure until 10 May 1972, when communications jamming figured in the destruction of one MiG-21 and seven MiG-17s. [27]

The Use of Chaff during Rolling Thunder

Unlike the complex self-protection pod, chaff, the simplest of radar countermeasures, saw limited service during Rolling Thunder. One reason was the Navy's concern that screens dense enough to blanket North Vietnamese radar would interfere with electronic equipment on board ships, located off the North Vietnamese coast, controlling fighter cover. Another reason was that Air Force fighter-bombers carrying the war to the North were not equipped with chaff dispensers.

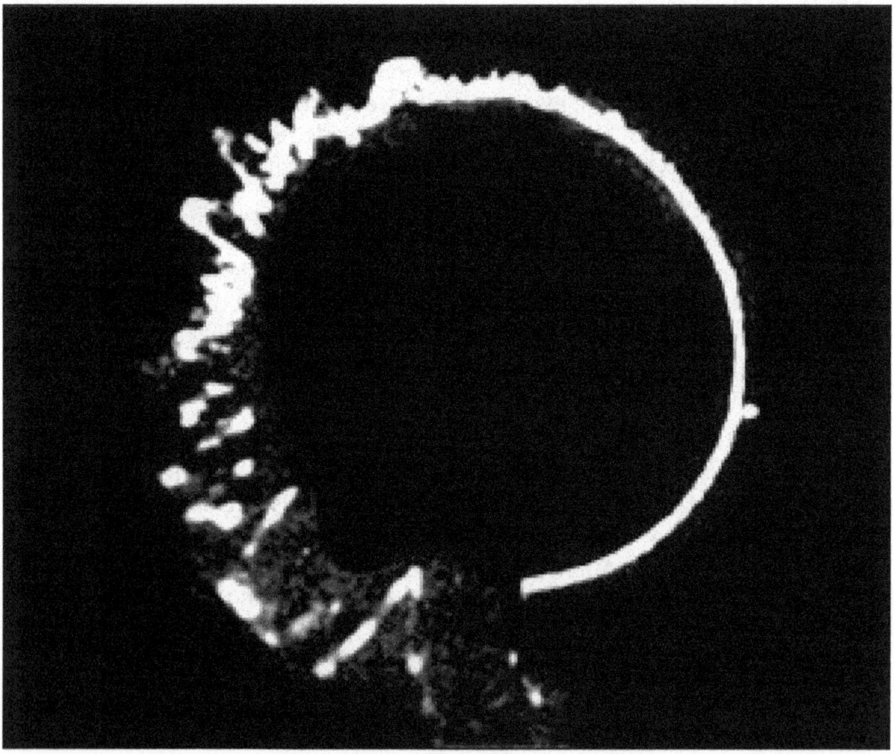

The effect of chaff on a radar screen. The right hand half of the circle shows an unaffected radar echo. The left hand half, particularly the lower left hand corner, shows the effect of chaff. What worries the operator is what might be coming at him with malignant intent through the disrupted coverage. Source: National Archives.

As early as the spring of 1967, however, F-4C squadrons had improvised a means of employing chaff in conjunction with other countermeasures. Bundles of the radar reflectors were taped inside the speed-brake well. When a SAM was sighted, or when the warning gear indicated that a missile was on the way, the pilot activated the brake, releasing a cloud of chaff, then took evasive action. The Udorn-based RF-4C reconnaissance unit proved equally inventive. When the pilot's warning equipment alerted him that radar had locked onto his plane, he used his photoflash cartridge dispenser, intended for night photography, to drop two or three chaff containers. [28]

While the EB-66's did carry chaff dispensers they used chaff only to supplement the electronic noise barrage. These planes released chaff to compensate for the weakening of the jamming signal when not flying broadside to the target radar. During the usual elongated orbit, each plane dumped chaff whenever it made a turn. The resulting chaff return posed no problem for shipboard fighter controllers, because the radar return was concentrated in a compact area. [29]

Self-Protection Pods: An Aid but not a Panacea

Comparison of Air Force losses in the same region, before and after pods became standard equipment, indicated that these devices reduced casualties, since the loss rate without pods was four time the rate with them. Although this comparison was far from conclusive because of the many variables involved -- specific target, weather, number of aircraft, and experience of both attackers and defenders -- pilots considered the pod essential to their survival, one of them crediting it with "turning the attrition rate upside down. " [30]

Arguably, the pods saved the F-105 fleet from destruction. Source: U.S. Air Force.

Despite its importance, the pod imposed some limitations on an attacker, drawbacks that stemmed mostly from the need to maintain formation in order to lay down an adequate jamming barrage. The pod also restricted maneuverability because steeply banked turns directed the strongest portion of the jamming signal ineffectually into space. Unfortunately, frequent turns were necessary to obtain an unobstructed view behind the formation and guard against attack from the rear, the favorite approach of MiG pilots. The fighter-bombers and their escorts tried, however, to compensate by making successive shallow turns, keeping the wings as level as possible and focusing the noise barrage downward. Another deficiency was that the pod increased drag and supplanted bombs or fuel that might otherwise hang beneath the wings. Without question, however, the protection provided was worth the sacrifice in maneuverability, fuel capacity, and bomb load. [31]

The pod also had some effect on bombing accuracy. By jamming the radars controlling the SAMs and 85-mm guns, it enabled the F-4s and F-105s to attack from above the concentrated fire of the automatic weapons and light antiaircraft guns which had claimed so many victims during 1965 and 1966. Bombing from these safer altitudes was generally considered less accurate than low-level strikes. Yet this loss of accuracy probably was theoretical rather than real, since low-altitude strikes could not achieve bombing- range precision in the face of intensive fire. [32]

V: ROLLING THUNDER TO LINEBACKER

A Time Of Transition

On 1 April 1968, President Johnson limited Rolling Thunder to targets south of 19 degrees North latitude; then, effective 1 November of that year, he halted the air war against North Vietnam. The latter decision did not put an end to all missions over the North, however, for both President Johnson and his successor, Richard M. Nixon, asserted the right to conduct aerial reconnaissance and to respond to enemy provocation.

Reconnaissance, whether by Teledyne Ryan drones or by manned aircraft, required countermeasures support, as did retaliatory strikes against carefully selected targets outside North Vietnam, such as airfields, and the supply depots that sustained ground operations in Laos, South Vietnam, or Cambodia. In attacking the enemy's lines of supply and reinforcement, B-52s ranged close enough to North Vietnamese territory to come within range of SAM batteries located there. Consequently Wild Weasels and EB-66s joined forces to protect the Stratofortresses from these weapons.

Protecting the B-52 Cells over Laos became a new priority. Source: U.S. Air Force

Despite some refinement, especially in Wild Weasel operations, jamming techniques during the bombing halt remained unchanged, even though Rolling

Thunder shifted to a smaller geographic area and then shut down. The principal change was an increased use of chaff to supplement stand-off jamming. As chaff screens became more common, the use of self-protection pods declined, for few of the retaliatory strikes hit the kind of heavily-defended targets where pods were most effective. Beginning in April 1968, therefore, Wild Weasel crews had to adjust their tactics when escorting armed reconnaissance flights, as they frequently did until the November bombing halt. Also, suppressing SAM sites that threatened B-52 cells required different operating methods than those used by the Wild Weasels during Rolling Thunder strikes.

Since resumption of all-out aerial war against the Northern heartland remained a possibility, Thailand-based Air Force squadrons occasionally rehearsed tactics used almost daily during Rolling Thunder. One such rehearsal occurred in May 1969, some 6 months after President Johnson halted the bombing. A formation of F-105Ds took off from Thailand heading toward the Red River delta. The formation turned back short of the North Vietnamese border, but not before triggering enemy radar. To add to the realism, an EB-66E and an EB-66C jammed early warning and ground control intercept sets, as had been done before the bombing halt. [1]

That "Old, Tired Airplane"

The EB-66 continued to be the only Air Force plane engaged in stand-off jamming. But the slower pace of air operations against the North required fewer of these aircraft, so EB-66 strength in Thailand diminished accordingly. On 31 October 1969, the 41st Tactical Electronic Warfare Squadron disbanded, From a maximum of 38 aircraft, the total in Southeast Asia now declined to 20 -- 6 EB-66Cs and 14 Es -- and the last of the B models headed for retirement at Davis-Monthan. [2]

By any standards, the EB-66E was reaching the end of its life by 1968. It, and most importantly, its engines were tired and spares were in short supply. Source: U.S. Air Force.

Some 12 months before the end of Rolling Thunder the enemy began shifting his defenses to deal with the B-52s attacking targets in the vicinity of the Ban Karai and Mu Gia passes into Laos, and in the area immediately south of the demilitarized zone. Just inside their own territory the North Vietnamese set up SAMs that could fire upon Stratofortresses as far as 15 nautical miles beyond the border. To counter this threat, EB-66s escorted the bombers and provided stand-off jamming.

The B-52 raids continued, unaffected by the ban on bombing the North and help from the EB-66s remained a necessity. On a typical mission, one or more EB-66s closed to about 10 nautical miles from the target, but remaining outside North Vietnamese airspace, and commenced jamming the radars located a short distance inside enemy territory. Electronic warfare officers on board the B-52s kept watch for hostile radar activity, turning on their own jammers to reinforce the EB-66 barrage should the North Vietnamese begin transmitting. [3]

During a December 1969 mission over Laos, SAMs located inside North Vietnam fired upon two EB-66s and the B-52s they were screening. Neither of the EB-66 crews detected a Fan Song signal, though the B-52 electronic warfare officers did. This indicated that at least one Fan Song operator used his radar to pinpoint the range a few seconds before launching, while others tracking the EB-66s relied solely on passive tracking in an apparent instance of triangulation. A SAM passed within 50 feet of one of the Stratofortresses, but neither of those fired at the EB-66s came closer than 5000 feet. [4]

Coverage provided by the SA-2 batteries was slowly expanding south and, as it did so, it pushed the EB-66s away from key areas. Source: People's Liberation Army of Vietnam

Even though Rolling Thunder had ended, Air Force EB-66s furnished jamming for the frequent drone reconnaissance missions over the northern provinces of North Vietnam. Electronic warfare officers engaged the radars that could locate the pilotless craft and direct SAMs, antiaircraft fire, or interceptors against them. From two to four EB-66s usually took part in this jamming, depending upon such factors as weather, the area to be reconnoitered, and its defenses.

When the drones were flying at low altitude into the SAM defenses protecting Hanoi and Haiphong, the EB-66s aligned themselves with the programmed flight path so that the most dangerous of the enemy's acquisition and missile system radars would be transmitting directly into the jamming beam. As the drone passed beyond the SAM sites, the Fan Song radars now looked away from the EB- 66 orbit and were all but unaffected by the jamming barrage. In contrast, the Spoon

Rest acquisition radar remained susceptible to jamming even though the noise source was behind it. Interference with Spoon Rest usually enabled the drone to survive the SAM defenses and escape to some lightly defended area of North Vietnam.

Until May 1969, the EB-66 bore exclusive responsibility for stand-off jamming to screen drone reconnaissance flights. During that month, however, the Marine Corps EA-6A began sharing the burden. Equipped with a steerable antenna, the Marine aircraft proved more successful than the older EB-66 in providing protection against North Vietnamese SAMs.

This AQM-34Q is a COMINT version of the basic Teledyne-Ryan AQM-34 Firebee drone. Launched in mid-air from a modified C-130, the AQM-34Q flew a preprogrammed course or was manually flown by a remote operator. It intercepted radio signals from as far as 300 miles away and relayed them in real time to a ground control van. After returning to a safe area over water, the AQM-34Q deployed a parachute. A modified helicopter then hooked the parachute to catch the drone in mid-air -- if the operation, failed the drone was retrieved from the ocean's surface. This particular AQM-34Q was nicknamed the Flying Submarine because of the many times it dropped into the ocean. Source: U.S. Air Force.

In addition to SAMs, the enemy sometimes used MiGs against the pilotless aircraft, and occasionally the two in combination. For about 3 weeks, the missile sites would challenge each drone entering heavily defended portions of the northern provinces. Then, for perhaps 2 or 3 weeks, MiGs took over while the SAM batteries remained generally silent. From the enemy's point of view, this pattern of reacting to the 20 to 55 monthly drone missions enabled him to exercise two components of the air defense system. Although valuable for training, this practice permitted the EB-66s and EA-6As to concentrate on just one threat; rarely did the Americans have to deal simultaneously with both missiles and interceptors.

Unlike the low altitude drones, those employed at high altitude carried equipment to counter both SAMs and MiGs. One item. Rivet Bounder, was a deception jammer activated by the Fan Song guidance signal. Because space was at a premium on board the reconnaissance vehicle, the same antenna that received the guidance signal also broadcast the false radar return. This resulted in a slower reaction than was normal for repeater jammers. In addition, high-altitude drones carried Hat Rack, a device that could recognize both MiG and SAM threats and induce an appropriate response. To frustrate a MiG attack, Hat Rack triggered a maneuver called "gimp". To counter a SAM, it could initiate two other evasive actions: "evade" and "dodge". The type of anti-SAM maneuver depended upon the power and direction of the Fan Song beam and whether or not a guidance signal was detected. Although neither Rivet Bounder nor Hat Rack was more than partially successful the high-altitude craft usually reconnoitered lightly-defended areas and therefore did not require the assistance of EB-66s or EA-6As.

The U.S. Air Force considered the Bar Lock radar to be a formidable adversary. Source: U.S. Air Force.

Several of the low-altitude drone routes employed around Hanoi and Haiphong exited to the southeast, and the pilotless craft neared the coastline, they usually climbed abruptly. Once the enemy noted this stereotyped flight profile, he directed MiGs against this portion of the routes. During the resulting attacks, the EB-66s and EA-6As focused upon the Barlock ground control intercept radar, but the jamming was generally ineffective against a set that the Air Force Electronic Warfare Center considered "excellent. " Occasionally the supporting aircraft also jammed very-high-frequency radio to disrupt instructions from ground controllers to interceptor pilots. Such communications jamming by an EA-6A probably saved a drone attacked by MiGs on 16 May 1970. Under almost the exact conditions, on 3 July 1970, the supporting EB-66s did not attempt to jam radio traffic, and the interceptors downed the reconnaissance craft.

The drones were usually launched from Navy or Air Force DC-130s. Source: U.S. Navy.

Participation by Marine Corps EA-6As in the drone program ended on 27 June 1970. Afterward, the low-altitude flights depended for survival on EB-66 support, and the high-altitude craft, protected by Rivet Bounder and Hat Rack, continued to fly over less dangerous areas without stand-off jamming. [5]

Following the 1 November 1968 bombing halt, EB-66 tactics tended to become stereotyped. When the aircraft worked with B-52s, for example, the defenders could be sure that the bombers would attack within SAM range of North Vietnam's borders, either along the demilitarized zone or in the vicinity of the mountain passes leading into Laos. Similarly, as an EB-66 crewman pointed out, "since so few EB-66s fly north over the Gulf of Tonkin, it is a good indication that when an EB-66 proceeds north . . . the drone launch is imminent. " In either case, the North Vietnamese had several clues to the identity of the countermeasures aircraft, such as the characteristic radar return, speed and altitude, and radio call signs, which were changed infrequently. [6]

Despite this tendency toward standardization, in part a consequence of brief tours and frequent rotation for both staff officers and flight crews, the Americans did occasionally vary their drone tactics by launching over Laos instead of the Tonkin Gulf. On 31 July 1969, one of these west-to-east flights caught the defenders off guard. And, while an EB-66C and EB-66E laid down a jamming barrage, a drone flew low over Hanoi, unchallenged by either gunfire or MiGs. [7]

The most serious problem confronting EB-66 crews during the late 1960's and early '70's was engine wear. In the words of Capt. James L. Hendrickson, an electronic warfare officer, the EB-66 had become "an old, tired airplane . . . really gone to the point where it's dangerous." [8] In April 1969, an Allison J-71 engine failed on takeoff, and an EB-66B crashed, killing all three crewmen. Inspection of the wreckage disclosed metal fatigue in the failed fourth stage compressor; as a result the entire EB-66 fleet had to be grounded. Mechanics pored over the Allison engines and discovered that fatigue cracks were common through the compressor stages of those engines with 1,200 or more hours of operating time since the last replacement of rotor discs. Slightly more than a third of the aircraft at Takhli were found to require engine changes, a job that took 2 months. Although the immediate crisis had ended by June, engine wear was a recurring problem for the remainder of the war. [9]

Wild Weasel in Quieter Times

From 1 April through 31 October 1968, bombing restrictions kept Air Force strikes south of the 19th parallel. The usual combat mission flown during this period was daylight armed reconnaissance. Because small formations ranged over large areas, the Wild Weasel tactics that had evolved for massed Rolling Thunder strikes now required revision.

Nor could the enemy mass his missile battalions as he had done when the Americans were concentrating upon targets in the Red River delta. Except for the SAM complex at Vinh, which resembled those that had been encountered farther north, missile units in the panhandle moved frequently and went into action only when a good target appeared. SAMs were least active along the demilitarized zone, where crews tended to look for easy kills.

In this new situation, Wild Weasels orbited in the vicinity of any probable or known missile units and attacked even the suspected sites during and after a strike. Because of the changed tactical situation, the mission was neither a radar suppression nor a hunter-killer activity, but a combination of both.

During April 1968, four SA-2 units were operating south of 19 degrees. Two of these defended Vinh, one was located along the demilitarized zone, and the fourth was sporadically active around Bai Due Than, a town mid-way between Vinh and Dong Hoi. The following month saw the addition of at least three units, one of which deployed to Vinh while another set up near Dong Hoi. The reinforced SAM defenses in the panhandle were not especially active, however, launching perhaps 10 missiles in May compared to 3 in April. During May, however, Wild Weasels undertook an aggressive bombing campaign against the Dong Hoi, Bai Due Than, and demilitarized zone missile defenses, making at least 26 attacks, in contrast to the one recorded for the previous month. The number of strikes against these SAM sites increased to 38 in June and reached a peak of 82 in August.

Although most of the bombs fell upon missile installations that appeared to be unoccupied, Wild Weasel crews did see many secondary explosions. Intelligence could not, however, determine how much SAM equipment was actually destroyed. Whatever the amount of damage, the strike may have discouraged the southward deployment of additional SAM units, for the build-up which had begun in May came to an abrupt end. By September, intelligence was finding it difficult to determine whether any missile sites remained in operation south of the 18 parallel. If such units were present. their activity had declined sharply from June through October they made just two single-missile attacks on American aircraft. The results attributed to Wild Weasels cost two F-105Fs downed by antiaircraft fire -- one in April and the other in September -- but no Air Force plane fell victim to SAMs. [10]

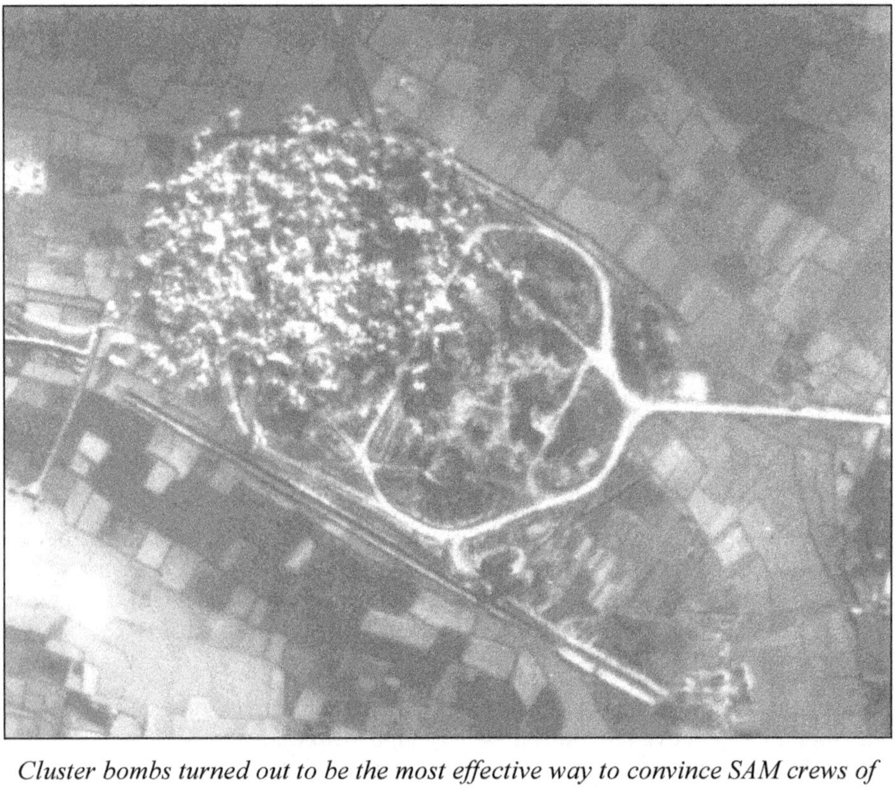

Cluster bombs turned out to be the most effective way to convince SAM crews of the error of their ways. Source: U.S. Air Force.

In addition to attacking missile sites, Wild Weasels were escorting manned reconnaissance planes over southern North Vietnam. The F-105Fs and Gs tried to suppress enemy radar by their presence alone. Indeed, for a time the rules of engagement specified that RF-4C or Wild Weasel crewmen actually had to see a missile hurtling skyward before the escort could attack a SAM site. This policy, however, was short lived, because crew members could not see missiles boring through low lying clouds in time to take evasive action. Moreover, the delay in reacting gave Fan Song operators time to shut down and thus deprive anti-radiation missiles of a target. [11]

While protecting the B-52s that continued to bomb infiltration and supply routes exiting from North Vietnam into Laos, Wild Weasel crews sometimes conducted trolling operations along the North Vietnamese border. According to Lt. Col. Robert E. Belli, a Wild Weasel pilot, the F-105Gs "would go along the borders of North Vietnam in the hope that radars would come up and we would have a chance to pinpoint any threat before the B-52s got there." When a Wild Weasel picked up a Fan Song signal, the crew warned the B-52s, which were usually under orders to avoid approaching an area defended by SAMs when one of these radars was transmitting. [12]

An F-105G flying cover for a B-52 formation. This aircraft carries an AGM-45 on the starboard wing and an AGM-78 on the port wing. Source: U.S. Air Force.

In protecting the B-52s, those Wild Weasels armed with AGM-45 Shrike missiles sometimes had to fly over North Vietnamese territory, where the SAM batteries and radars were located. "We felt, " Lieutenant Colonel Belli reported, "that we had to position ourselves between the B- 52 strike force and the nearest SAM site." Belli and the other crewmen wanted the enemy to see the Wild Weasel escort and realize the danger of using radar-controlled weapons against the bombers. However, because of the short range of the AGM-45, "many times this meant we had to be over North Vietnam. " Otherwise, Shrike launchings would be futile and radar suppression a joke. Whether or not to enter North Vietnamese airspace "was pretty much our interpretation of when it was required. " [13]

Usually two Wild Weasels supported each B-52 mission, flying parallel to the border and timing their runs so that the Stratofortresses had coverage during approach, bomb release, and withdrawal. As the ordnance was exploding among hidden trails, and storage areas, and the bombers began turning away, one of the F-105Gs generally flew straight toward the SAM site that posed the greatest danger, maintaining course for 30 to 60 seconds in order to ensure suppression until the B-52s were safely out of range. [14]

One of the two Wild Weasels escorting a B-52 strike normally carried a Standard ARM AGM-78 missile and a Shrike, and the other a pair of Shrikes. Most F-105G crews preferred the AGM-78, even though its weight and aerodynamic characteristics reduced aircraft performance Lieutenant Colonel Belli maintained that the standard ARM "gave us a real long range capability against the SAM plus, in certain scenarios, our missile could impact on a SAM site before their missiles could hit us, just as a function of speed. " He felt that the Shrike remained in service only because of a shortage of the fittings used to secure the newer missile beneath the wings of F-105Gs. [15]

Even though North Vietnam's outermost defenses were less formidable than those encountered by the Wild Weasels during Rolling Thunder, radar suppression was still dangerous work. Lieutenant Colonel Belli, while still a major, and his electronic warfare officer, Lt. Col. Scott W. Mcintire, found themselves flying above a solid cloud deck on 10 December 1971 while supporting a B-52 strike in Laos near the North Vietnam border. Their homing and warning set told them a SAM was being launched, but the clouds prevented them from seeing the missile in time to avoid being hit. Both men ejected from the burning plane, but Belli alone survived, to be rescued the following day. [16]

Out of the Eclipse: Self-protection Pods in the Son Tay Raid

Since most American air operations by 1971 were taking place along the periphery of North Vietnam instead of in the more heavily defended heartland, self-protection pods were no longer as essential as they had been during Rolling Thunder. For example, Wild Weasel crews sometimes flew F-105G's that did not carry the special QRC-335 jammers developed for them. In fact, Lieutenant Colonel Belli was flying such an unprotected airplane when he was shot down in December 1971. He maintained that the eclipse of the self-protection pod was "one example of experience and lessons learned and forgotten during a period of time because the turnover [of personnel] was so heavy and the threat so light. " [17]

Pods did prove invaluable, however, when Air Force planes had to penetrate deep into the North, as happened during November 1970 when a raid on the village of San Tay attempted to free 50 to 60 of the estimated 350 American servicemen held prisoner in North Vietnam. Preparations for the raid began in the summer of that year. A joint task force, commanded by Brig. Gen. LeRoy J. Manor, USAF, planned and trained for a swift descent upon Son Tay, some 20 nautical miles northwest of Hanoi and the site of a compound that, according to intelligence data, held as many as 60 prisoners of war.

The Son Tay Camp. For some reason, the buildings in this picture are labeled in Spanish. Source: U.S. Air Force

Brigadier General Manor and his staff planned for an Army assault force, carried in Air Force helicopters, to rescue the men. Fighter cover and air support were Air Force responsibilities, and Navy carrier pilots were to conduct a feint to rivet the enemy's attention to the port of Haiphong 60 nautical miles southeast of the actual objective. Rehearsals were held in the United States using a full-scale, wood-and-canvas mockup of the prison camp, and, after a final briefing in Southeast Asia, on the night of 20-21 November the group headed for Son Tay.

While 59 Navy planes executed a diversion off Haiphong, 6 helicopters flew eastward from Thailand, following their navigation ship, a Lockheed C-130E Hercules. Another Hercules led the strike force, five Douglas A-1E attack planes furnished fire support. Both transports mounted jamming transmitters to frustrate gun-laying radars like Fire Can, and the A-1Es carried QRC-128 jammers to disrupt radio communication between MiG pilots and their ground controllers.

Also on hand that night was a support group which included Air Force F-4 Phantoms for fighter cover, a Lockheed EC-121 to direct these fighters and provide MiG warning, a Boeing RC-135M for communications intelligence, and a KC-135 airborne radio relay. A pair of HC-130P tankers was available to refuel the helicopters, while KC-135's were on hand to refuel the F-4s and five Wild Weasel F-105Gs included in the expedition.

These Wild Weasels, armed with anti-radiation missiles, were to protect the assault force and its fighter cover from the SAMs that guarded the Son Tay area. Enemy missile crews managed, however, to damage two of the F-105Gs, one so seriously that it had to be abandoned over Laos. Both crew members parachuted safely, and were picked up by helicopter.

Within the prison compound, the raiders found only empty cells and some North Vietnamese troops who when they recovered from their initial surprise, opened fire. Despite this opposition, only one American was injured and none killed, though a disabled helicopter had to be destroyed as the force was withdrawing.

Although the basic plan worked almost flawlessly. The operation failed to accomplish its purpose. Sometime during the autumn the enemy had transferred the prisoners from Son Tay to other facilities. [18]

Rehearsing for the Final Act

The countermeasures gear that had served so effectively in Rolling Thunder was returned to action during retaliatory operations like Proud Deep Alpha -- a series of strikes against stoutly defended airfields, petroleum storage tanks, and military barracks. These attacks, delivered from 26th through 30 December 1971, represented a response to North Vietnam's shifting of MiG interceptors and additional SAMs to protect Ban Karai and Mu Gia passes, two portals for the movement of troops and supplies from Laos to South Vietnam. During the 5-day operation, Air Force and Navy planes flew 935 strike sorties, defying cloud cover to bomb targets located in the North Vietnamese panhandle between the 17th and 19th parallels.

An additional 29 sorties were devoted to armed reconnaissance, all but impossible because of the foul weather, and 102 to radar suppression, a particularly dangerous mission because the low-hanging clouds that concealed enemy SAMs and gave the Iron Hand crews just moments for evasive action. [19]

Although no MiGs rose to defend the airfields or other Proud Deep Alpha targets, Air Force and Navy fliers met determined opposition from SAMs and antiaircraft guns. Because of the clouds, American airmen may not have seen all the missiles launched at them, but 45 SAMs were reported plus two "possibles," with 22 of the missiles having been fired on a single day, 30 December. Two Navy aircraft, an F-4 and an A-6A, and an Air Force F-4 fell victim to SAMs, during the operation. [20]

Iron Hand flights escorted Proud Deep Alpha strikes against installations protected by SAMs, attempting to suppress Fan Song radars much as they had done during Rolling Thunder. For the Wild Weasel role, the Air Force again relied upon F-105Gs armed with Shrike and Standard ARM missiles, plus 20-mm cannon, but without the bombs formerly carried. The Navy used A-6As and Ling-Temco-Vought A-7Es, most of them able to launch either of the two anti-radiation missiles. These Navy planes carried basically the same radar homing and warning equipment as the Air Force Wild Weasels. Although Fan Song was the primary concern, the F-105Gs and their Navy counterparts sometimes tried to suppress gun-laying, and ground control intercept radar. [21]

Determining Wild Weasel effectiveness during Proud Deep Alpha proved even more difficult than during Rolling Thunder. Bad weather complicated the evaluation, a task already made difficult by enemy radar technique. Shortly after Shrike launchings became commonplace, the North Vietnamese had learned to detect the launch of an anti-radiation missile and cease transmitting before the weapon could home on the radiation source. Once they discovered that Standard ARM had a memory circuit to keep it on target, they modified these tactics. Whenever possible, a second radar of similar type began transmitting as soon as the probable target shut down, remaining on the air just long enough to divert the AGM-78 from its intended victim.

The fact that a radar was listed as only "probably destroyed" even though no further signal emanated from the site for 2 weeks. These considerations helped account for an unimpressive damage estimate that credited 51 Shrikes and 10 AGM-78s with one ground control intercept radar "killed," plus two other acquisition radars, five Fan Songs, and four Fire Can or similar radars "possibly destroyed." [22]

The ZU-23-2 was a new arrival that made low-altitude attacks even more dangerous. Source: U.S. Army

Participants in Proud Deep Alpha encountered moderate to heavy fire from every antiaircraft weapon in North Vietnam's arsenal, from machine guns and newly arrived 23-mm cannon, through 37-mm and 57-mm, to 85-mm and 100-mm guns. The lighter weapons fired planned barrages, while most of the heavier ones were radar controlled. Although American crews reported seeing a total of 800 shellbursts during the 5 days, this fire did not destroy a single Air Force plane. This lack of effectiveness probably resulted from electronic countermeasures in the case of radar-controlled weapons, and from an overcast that neutralized optical aiming devices. [23]

The EB-66, relegated to a secondary role by the time Rolling Thunder ended, figured prominently in the Proud Deep Alpha countermeasures effort. For the first time in 5 years, the plane ventured within range of SAM batteries, relying on Iron Hand radar suppression and its own jamming power to avoid destruction. This was risky, as Lt. Col Frank R. Wink, USAF and his crew could testify. On 29 December, a SAM site protecting Quan Lang airfield fired at Wink's EB-66E. The electronic warfare officer detected a Fan Song radar in the high pulse repetition frequency and then the launch signal. He warned the pilot, who saw the missile in time to make a diving turn to the left. As this SAM was exploding harmlessly high above the plane, the electronic warfare officer reported another launch and moments later a third. Wink eluded both missiles and wasted no time getting clear of the battery that had launched the salvo.

On the same day, another EB-66E pilot, Lt. Col Jack E. Tullet, led a three plane element deep inside North Vietnam. Tullet's EB-66E and another plane of the same type jammed acquisition radars and helped a force of 34 Phantoms bomb their targets without losing a single plane. The third plane in Tullet's flight, an EB-66C, tuned in on Fan Song frequencies and succeeded in pinpointing every one of these radars transmitting from the vicinity of Mu Gia Pass. [24]

Because of the cloud cover that reduced bombing accuracy, Proud Deep Alpha attained only partial success in terms of damage inflicted, with some 88 percent of the bombs dropped detonating within the target areas. As far as electronic countermeasures were concerned, this operation, like the other retaliatory strikes that preceded it, polished rusted skills and reacquainted air crews with the equipment they would need for attacks in regions skillfully defended with radar-controlled weapons. Quite by accident, Proud Deep Alpha served as a rehearsal for the resumption of the air war against the North in response to the invasion of South Vietnam in the spring of 1972. [25]

VI. THE LAST AIR BATTLES

The ground war in South Vietnam remained generally stable following the 1968 bombing halt. On 30 March 1972, however, this battlefield stability vanished amid exploding artillery shells as North Vietnamese troops invaded the South. President Nixon reacted by dispatching additional planes to Southeast Asia and inaugurating an aerial campaign against the North that quickly surpassed Rolling Thunder in intensity. The new air effort had the same purposes as the old: to pressure North Vietnamese leaders into halting their aggression, and to impede the movement of enemy troops and war material into South Vietnam.

The 1972 air war against the North began on 6 April with Operation Freedom Train, a 1-month campaign south of the 20th parallel. The war moved north of that line for Operation Freedom Porch Bravo, a series of strikes in the Hanoi-Haiphong region delivered on 16 April. Then Linebacker, which evolved from Freedom Train, got underway on 9 May. This latest operation attempted to sever North Vietnam's overland transportation arteries, particularly the rail line to China. To complement the strikes on roads and railways, President Nixon approved a blockade of North Vietnam that included sowing aerial mines in the harbors and river mouths where Russian military supplies were arriving. This escalation of the air war reflected the mounting pressure being exerted upon South Vietnam's defenses by the invading North Vietnamese. [1]

When Miss Buffy bombed something, she really bombed it. Source: U.S. Air Force

The renewed aerial campaign included frequent B-52 strikes against the North. The early B-52 attacks were: Freedom Train Bravo (Vinh, 9 April); Freedom Dawn (Bai Thuong airfield, north- west of Thanh Hoa, 12 April); Freedom Porch (Haiphong, 15 April); Freighter Captain (Thanh Hoa, 21 April); and Frequent Winner (Thanh Hoa, 23 April). On 9 April, these aircraft made their deepest thrust of the war into North Vietnamese territory, when a dozen Stratofortresses attacked the petroleum storage tanks and rail yard at Vinh. Three days late the big bombers struck further north, pounding Bai Thuong airfield, a MiG base just 6 nautical miles south of the 20th parallel. The following week, they made their first foray into the most heavily defended region of North Vietnam, a 15 April attack upon a petroleum storage depot at Haiphong.

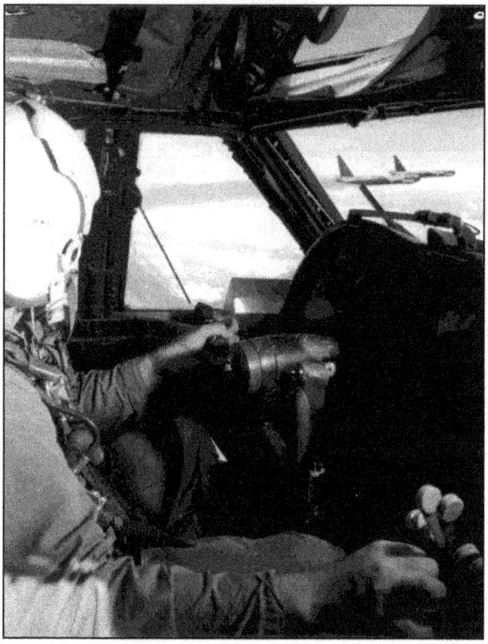

A B-52 cell viewed from the cockpit of one participant. Source: U.S. Air Force.

Following the Haiphong strike, the B-52s shifted their effort south of the 20th parallel, making two attacks on Thanh Hoa before the end of April. By this time, Thanh Hoa's defenders had become proficient at tracking the jamming source. On the 21st they launched 11 SAMs, 4 of them at B-52s that were turning after releasing their bombs, but inflicted no damage. Two days later at least a dozen missiles roared skyward, all of them aimed at the jamming source. One SAM, out of a salvo of four directed against a three-plane cell, exploded beneath a bomber that was approaching the target. Despite damage from fragments, the aircraft landed safely at Da Nang. The B-52s took no evasive action during the Thanh Hoa raids, relying instead upon chaff, Wild Weasel escort, and electronic jamming to thwart the defenses. The fact that only one bomber sustained damage testified to the effectiveness of the countermeasures. Following the Thanh Hoa strikes, the B-52s concentrated on southern North Vietnam. By the end of November, they had made more than 2,300 sorties against targets in this area, hammering the supply lines that sustained the invasion force. Meanwhile, on 23 October, Mr. Nixon announced a suspension of the bombing north of the 20th parallel banning, at least temporarily, attacks on North Vietnam's heartland. [2]

The President restricted the bombing in the hope of encouraging a negotiated settlement. When diplomatic progress failed to materialize, however, he again authorized air attacks throughout all the country, except for a narrow buffer zone along the Chinese border. B-52s spearheaded this new offensive, Linebacker II, which began on 18 December and lasted until the 30th, when Mr. Nixon reinvoked the ban on bombing north of the 20th parallel. Strikes in the panhandle continued through 15 January, when the President halted all air and naval operations against North Vietnam. [3]

"Get All the Pods You Can Get Your Hands On."

When American fighter-bombers resumed their attacks throughout the North in the Spring of 1972, the self-protection pod regained the importance it had gradually lost after Rolling Thunder. Stressing the value of the pods, Lt. Gen. Robert J. Dixon, Vice Commander of the Seventh Air Force during the 1969-1970 bombing lull, pointed out that retaliatory strikes during the lull "weren't a fair test of the pods, " since planners "picked out isolated SAM areas and attacked them," avoiding any "real, intricate, pod-type problem." But, the general believed, in any future campaign in which "you've got to go to Hanoi, you'd better get all the pods you can get your hands on." [4]

During the 1972 fighting four basic types of self-protection pod were available, plus the jamming devices designed for Wild Weasel F-105Gs and General Dynamics F-111As. These four were: the ALQ-71 and ALQ-87, both veterans of Rolling Thunder; their designated successor, the ALQ-119; and the ALQ-101. From Freedom Train through Linebacker II, these pods performed both "normal" and "special" jamming, primarily at radar components of the SAM system.

The ALQ-119 pod on an F-4E. Source: U.S. Air Force

Normal jamming sought to disrupt the Fan Song tracking beam. To accomplish this, all four pods generated a modulated noise barrage, while the ALQ-119 and ALQ-101, in addition, could transmit a deceptive radar return. The basic noise barrage eroded Fan Song's ability to determine range, while noise modulation disrupted tracking. The deceptive signal also caused tracking error, though it had no effect on range calculation. When bombarded by all three kinds of normal jamming -- sustained noise, supplemented by noise modulation and deception -- the Fan Song operator was forced to resort to optical tracking, which was immune to electronic interference.

The ALQ-101 EW pod. Source: U.S. Air Force.

Special jamming was merely a new name for downlink or beacon jamming. Pioneered with the ALQ-71 and ALQ-87 during Rolling Thunder, special jamming consisted of modulated noise, broadcast on the frequency used by the SAM tracking beacon. This jamming signal caused the Fan Song trough antennas to pick up a distorted position signal to feed into the fire control computer, which then compared it to the desired trajectory and issued an erroneous guidance command. Moreover, the distorted position signal frustrated attempts by Fan Song crews to locate aircraft accurately enough to guide the SAMs optically. Linebacker fighter-bomber crews considered this special jamming technique their "single most effective ECM [Electronic Countermeasure]" and the "only ECM effective against the optically guided SAM." [5]

Technicians at the Tactical Air Warfare Center at Eglin AFB, Florida, raised the possibility that the ALQ-71, while jamming the down link, might be offering the Fan Song operator a well defined target for passive tracking. This concern, which did not involve either the ALQ-87 or ALQ-119, proved groundless. Extensive tests disclosed that the noise-modulation feature did highlight the aircraft, but at intervals so irregular as to defy tracking by the ablest of radarmen. [6]

Plans to replace the ALQ-71 and ALQ-87 with the ALQ-119 received a setback during the 1972 air war. When mounted in place of a Sparrow air-to-air missile on a fully laden F-4, the ALQ-119 had interfered with radar homing and warning gear, causing strobes of light on the viewer and a ringing in the earphones. Tests proved that the false signal appeared when bombs slung beneath the fuselage reflected the jamming beam. But, since the bombs could not be carried elsewhere, the easiest solution was to shelve the ALQ-119 in favor of the older pods. [7]

Interference also occurred when F-4s carrying television guided bombs tried to use their self-protection pods. The jamming noise so distorted the television image that accurate guidance was impossible. Technicians in the combat zone hurriedly installed makeshift screening and grounding to reduce the distortion, while the manufacturer provided permanent shielding for the electronic components of guided weapons leaving his production line. [8]

The F-111A already had a capable EW suite but it needed augmentation, proving the old adage, "if some is good, more is better." Source: U.S. Air Force

The General Dynamics F-111A figured prominently in the 1972 fighting, especially during Linebacker II, for night strikes on airfields and missile sites protecting Hanoi and Haiphong. Built into this fighter-bomber was an ALQ-94 deception jammer capable of frustrating Fan Song or gun-laying radar but ineffective against the SAM down link. The addition of an ALQ-87 compensated for this weakness, so by mid-August all F-111As were carrying the supplementary pod. [9]

F-111A crewmen considered the combination of ALQ-94 and ALQ-87 "sufficient to counter the enemy threat." But, actually, Linebacker placed few demands on the pods, for a combination of high speed and low altitude seemed to provide protection enough. In action, the planes usually attacked individually, skimming the ground at high speed. This approach apparently frustrated electronic tracking) judging from the tendency of radar-controlled antiaircraft batteries to resort to planned barrages. The gunners consistently underestimated the plane's speed, apparently firing at the sound of the twin turbo-fan engines, for the shells usually burst far astern, and it was not until late October that the enemy managed to place a barrage in the path of an F-111A. In such circumstances, the fighter- bomber crews refrained from jamming unless their warning equipment showed that a SAM or gun-laying radar was tracking them. These tactics proved successful, for, during the 3 weeks ending on 20 October, only 10 SAMs were launched at the F-111As and none scored a hit. [10]

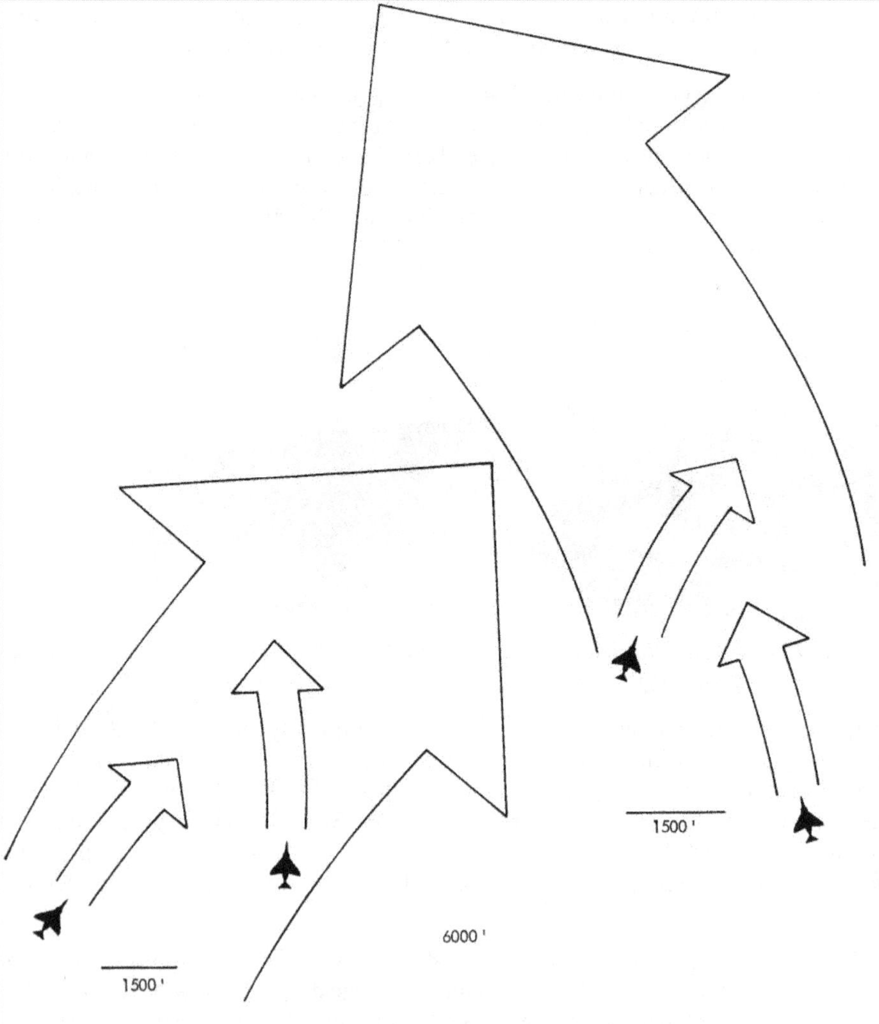

MANEUVER POD FORMATION

SCHEMATIC

The pod formations that served so well during Rolling Thunder did not however meet Linebacker needs. The earlier formations were a compromise between jamming effectiveness and bombing accuracy, but early in Linebacker the vulnerability of this compromise was exposed when during, May and June, enemy interceptors shot down 15 American planes. The weakness in the old pod formations was lack of visibility to the rear, toward 6 o'clock, the direction from which MIG pilots liked to attack, usually with heat-seeking missiles. To counter these tactics, Air Force tactical fighter wings adopted variations of a "maneuver" pod formation in which the two elements in the four-plane flight were separated laterally by as much as 6,000 feet, with about 1,500 feet between the planes in each element. Both elements weaved back and forth, the wing-men crossing and recrossing behind the element leaders, one of whom also led the flight. This continuous maneuvering enabled the wingmen to watch the danger area at 6 o'clock. In some units, the wingmen veered sharply when changing heading, with countermeasures responsibility devolving upon the element leaders, who made gradual turns, thus avoiding the steep banks that would divert the jamming signal. [11]

Early in Linebacker, all the tactical fighter wings adopted some variant of this maneuver pod formation for both the combat air patrol and the strike force. The new formation offered better coverage of 6 o'clock than did the old ones, and laid down an equally effective jamming blanket. Having "all aircraft . . . continually change their distance out, angle off, and altitude differential slightly" caused Fan Song scopes to reflect both "range and angular ambiguity." The weave that was devised to cope with MiG attack thus paid dividends in SAM protection. [12]

The new pod formations, however, like the old, were not ideally suited for dive bombing. The need to change heading and adjust position interfered with the task of locating camouflaged and heavily defended targets. To improve accuracy, pilots readopted the old Rolling Thunder practice of abandoning the pod formation about 1 minute from the target and rolling in individually. [13]

Horizontal bombing through clouds also posed a problem. The maneuver pod alignment resulted in a widely scattered bombing pattern when strike aircraft used LORAN, a navigation aid, to verify a release point. In order to obtain a satisfactory pattern from the usual bombing altitude of 20,000 feet, LORAN flights adopted a procedure whereby "the number 2 and 3 aircraft initiate a turn of 3 degrees heading into the flight lead as he calls 10 seconds. . . . [and the] number 4 aircraft initiates a turn of 6 degrees into the lead heading at 10 seconds prior to bomb release. " In effect, the maneuver reduced lateral separation between air-craft from 1,500 to 1,200 feet, while keeping "adequate individual separation in the event of a SAM detonation. " The planes released their bombs on signal from the flight leader. [14]

Assessing the effectiveness of the pods continued to be difficult. Statistics showed that American airmen sighted some 2,661 SAMs from April through October 1972 and that these weapons downed 41 pod-equipped planes, for a ratio of 65 launchings for every aircraft destroyed. During the same 7 months of 1966. 548 missiles shot down 15 planes, none of which carried pods, for a ratio of 37 to 1.

This would seem to indicate that the pod-equipped fighter-bomber was almost twice as safe as the plane that had none. Yet, statistical evidence was at best inconsistent, for the 1972 ratio of 65 to 1 was less favorable than that posted for the same period in 1967, when the enemy had to launch about 83 SAMs for every pod-carrying aircraft destroyed. Differences in weather, types and numbers of defensive weapons, the skill of the defenders, and ability to maintain a pod formation combined to subvert statistical analysis. [15]

Chaff: Essential for Survival

By 1972, tools for employing chaff were much improved over those used during Rolling Thunder when planes dumped chaff bundles from speed brake wells and flare dispensers. For Linebacker, the F-4s used externally mounted chaff dispensers to lay corridors rapidly and to broadcast leaflet bombs packed with chaff that burst at designated altitudes and quickly increased the depth and density of the protective screen. But even when the bombs released puffs of chaff as planned, the wind could prevent the chaff clouds from merging to create a corridor. [16]

Chaff (and flare) dispensers became ubiquitous as the threat from radar-guided weapons grew. These are installed on a C-130 but are typical of their kind. Source: U.S. Air Force.

On 10 May 1972, the 8th Tactical Fighter Wing severed the highway and rail spans of the Paul Doumer bridge at Hanoi, using laser and television-guided bombs. The attackers approached through a chaff corridor 2 nautical miles wide, 4000 feet deep, and 34 nautical miles long, laid by two four-plane flights of F-4s a quarter hour beforehand. Strike aircraft armed with the guided bombs used their airborne radar to remain within the corridor, or slightly above it, as they drew near the heavily defended bridge. To prevent the chaff from interfering with the bomb guidance system, the Phantoms dived below the corridor, releasing their ordnance from as low as 8000 feet. Although the enemy fired 160 SAMs during the attack, the missiles did no damage, thanks to the combination of chaff and ALQ-87 jamming pods. [17]

Similar tactics evolved for screening B-52 raid A total of 20 F-4s -- five four-plane flights in a maneuver pod formation -- dropped 120 chaff bombs from altitudes between 34,000 and 36,000 feet to establish a protective corridor for the Stratofortresses. Within the individual chaff-dispensing flights, aircraft were separated laterally by 1,000 to 1,500 feet, but the altitude difference between the leader and the highest plane was no more than 500 feet. The five flights, with 1 nautical mile lateral distance between them, arranged themselves in echelon right or left. At a designated point, the 01 and 02 aircraft of each flight dropped the first of their six chaff bombs, afterward releasing the others at 20-second intervals. Numbers 03 and 04 then took over, dropping their bombs in the same manner. With 10 of the 120 chaff born bursting at intervals of about 3 nautical miles, the planes fashioned a corridor 6 to 8 nautical miles wide, 30 to 35 nautical miles long, and about 4,000 feet deep.

At first, strike planners allowed just 15 minutes for the corridor to form, but experience showed that 20 to 30 minutes were needed. The Stratofortresses then arrived, usually attacking from 35,000 or 36,000 feet, beyond reach of 85-mm guns. Because an F-4 burdened with two self-protection pods and six chaff bombs could not claw its way above 36,000 feet, the B-52s had to remain above the corridor. By flying lower, the bombers could have passed through the heaviest concentration of chaff, where protection was greatest but the risk from antiaircraft barrages seemed too great.

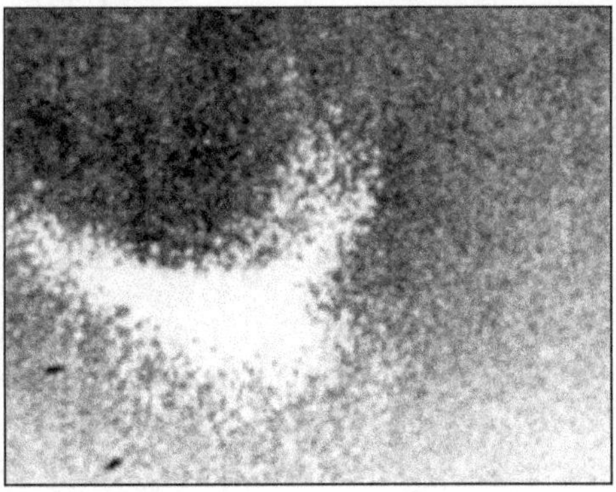

Photographs of chaff clouds are rare since people in situations that demanded dropping chaff usually had better things to do than take pictures. Source: Royal Air Force.

These strike tactics met a severe test on 15 April, when 17 B-52s braved the defenses of Haiphong to bomb a petroleum storage depot. The bombers followed a corridor formed by 119 chaff bombs, but an unfavorable wind prevented the reflectors from spreading as planned. On radar, the effect resembled a series of individual clouds rather than a corridor, but other countermeasures prevented the enemy from taking advantage of the gaps in chaff protection. Stand-off jamming by three EB-66Es, three EKA-3Bs, and an EA-6A -- plus the electronic barrage laid down by the B-52s -- kept the enemy from scoring hits on the Stratofortresses. [19]

As early as the summer of 1972, Air Force tactical units had a chaff dispenser capable of preventing the kind of gaps that had appeared over Haiphong. Planes fitted with the new equipment used it to reinforce corridors created by a series of chaff bombs, but experience showed that the corridors had tactical disadvantages. Besides being easily scattered by high winds, the chaff column alerted the defenders to the route the attackers would follow. An obvious solution was to saturate the target area with chaff, a technique for which the dispensers were well suited. But even so, area coverage did not emerge as the standard method of chaff usage until Linebacker II was well underway. [20]

Chaff-laying equipment reduced the speed and maneuverability, as well as the ceiling, of the F-4s. When burdened with a half-dozen chaff bombs a Phantom could not exceed 480 knots. Consequently the chaff flight, preceding the strike force by at least 15 minutes, presented an easy target for MiGs attacking from behind at supersonic speed with heat-seeking missiles. Two methods were devised for protection against the MiGs -- deception, and fighter cover. The former seemed to help, but the latter was effective beyond question.

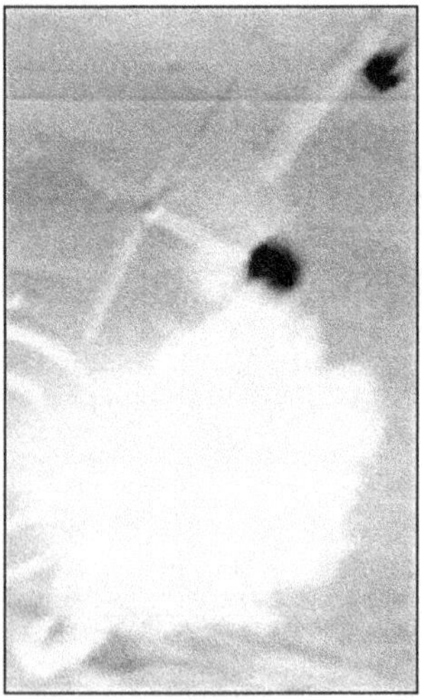

Shooting an enemy fighter – here a MiG-17 -- down is a good form of protection. Source: U.S. Air Force

One form of deception consisted of trying to persuade the enemy that the chaff flight was part of the combat air patrol usually by assigning radio call signs normally used by the fighter escort. Another was to vary the time interval between release of the first chaff bomb and arrival of the strike force. A third form of deception, having the chaff dispensers lay the corridor while flying toward the approaching fighter-bombers instead of leading the way to the target, was considered, but was abandoned as impractical because of the coordination required. [21]

Fighter protection functioned in either of two ways. In one the escort arrived on station as the chaff force commenced laying the corridor, or it accompanied the chaff-laden Phantoms as they approached the target, remaining with them throughout the creation of the 30-to-35-nautical-mile lane. In the second option, the escorting F-4s made frequent turns to keep from outrunning the chaff force, which was itself slowed both by the chaff bombs and by the need to weave back and forth in a maneuver pod formation. In making the successive turns, the escort also gained a tactical advantage, for when changing direction, the Phantom crews got an unobstructed view of the skies behind the chaff flight. Protection did not necessarily require the destruction of enemy planes, for an air-to-air missile from a patrolling F-4, even though launched at a poor angle, frequent persuaded a MiG pilot to break off his hit- and-run attack. [22]

Although chaff-burdened F-4s were encumbered, they were not defenseless, as was demonstrated on 15 August by a crew from the 8th Tactical Fighter Wing. Using tactics that had already claimed two of the wing's chaff carriers on earlier missions, a MiG-21 pounced on an F-4 flown by Capt. Fred Scheffler. The North Vietnamese pilot made a sloppy pass and overshot his intended victim. He may have thought that Scheffler's Phantom did not carry missiles; if so, it was a fatal mistake, for Capt. Mark Massen, in the back seat of the F-4, launched a Sparrow that exploded in the MiGs tailpipe. [23]

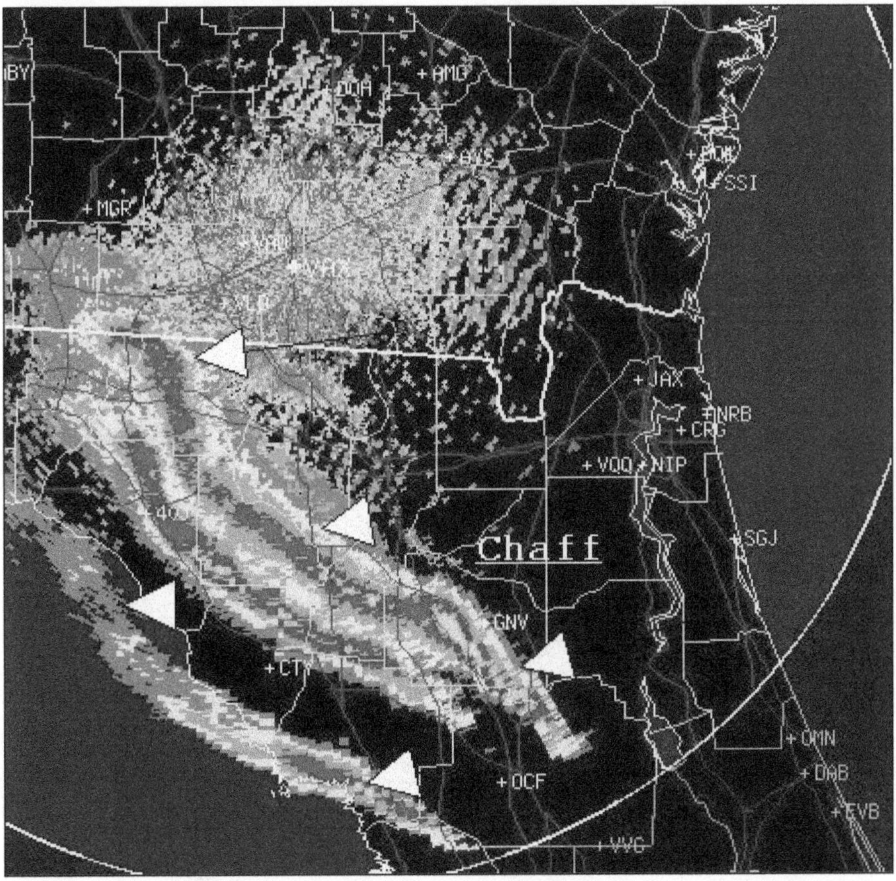

This picture, taken over Florida by the National Weather Service clearly shows the corridors of chaff that effectively obscure large areas of the state for the duration of the military exercise in progress. Source: National Weather Service, Jacksonville, Florida.

During the 1972 fighting, chaff earned the reputation of being "essential for the survival of the strike force in an SA- 2 and radar directed AAA environment, " for it affected both Fan Song performance and the functioning of the radar proximity fuse on the Guideline missile. [24] Although Fan Song operators could screen out chaff return accuracy suffered, so the chaff still accomplished its purpose. [25]

While laying the protective screen, however, the chaff flight itself was in danger from SAMs and heavy anti-aircraft guns. The chaff aircraft received no protection from the radar reflectors billowing out behind them; indeed, the trailing chaff pointed out the dispensing planes. Against radar-controlled weapons, therefore, chaff flights had to depend upon their own self-protection pods and on the Iron Hand aircraft operating far below them.

Certain types of friendly radar also picked up echoes from chaff. Strike aircraft, for example, could rely on their airborne radar to locate and follow the corridor. Return from the chaff did not prevent horizontal radar bombing, however, for during Linebacker II the B-52s located offset aiming points despite dense concentrations. And, when releasing electronically guided bombs, F-4s avoided interference by diving beneath the chaff corridor. [26]

During Rolling Thunder, the Navy had warned that heavy concentrations of chaff would interfere with equipment on board its command and control ship off the coast of North Vietnam. For Linebacker, therefore, a new control center, nicknamed Teaball was created in Thailand. This installation was apparently less sensitive to chaff, for despite the dense sown corridors Teaball controllers were able to successfully track MiG interceptors. The new control center was so successful that Gen. John W. Vogt, Jr., Commander, Seventh Air Force, paid tribute to its steady performance as a "sophisticated system for warning our guys." [27]

With chaff as with other countermeasures, analysts attempted to evaluate effectiveness statistically, sometimes with meaningless results. For example, the Seventh Air Force operations directorate reported that from 1 April through 31 August 1972, 13 Air Force planes fell victim to 1,325 SA actually sighted, a ratio of 102 missiles for each aircraft downed. In contrast, 983 SAMs destroyed 17 aircraft from August through December 1967, a ratio of 58 to 1. Citing these figures, the directorate declared that the "only major change" between the two periods was "the use of chaff," a false premise that invalidated the conclusion that chaff was the most successful of countermeasures. [28] As the Air Force Electronic Warfare Center pointed out, this kind of oversimplification could lead to "erroneous conclusions about chaff effectiveness." [29]

A study done for the Joint Chiefs of Staff avoided comparisons between Linebacker and Rolling Thunder but focused instead on the air war from April through October 1972. This study disclosed that Navy pilots, who used chaff mainly for last-minute self protection, reported an average miss distance for SAMs of 1,586 feet when chaff was released, compared to 1,162 feet when it was not. Air Force fighter-bomber crews who followed chaff corridors rather than releasing puffs of chaff when already in danger, reported an average miss distance 1,000 feet greater than that experienced by Navy fliers. SAMs, moreover, hit only two aircraft concealed in the corridors, one of which managed to limp safely home to bolster the argument in favor of the lavish use of chaff. The JCS study pointed out that 22.5 percent of the SAMs which missed planes outside the chaff screen passed within 200 feet of the intended victim; inside the corridor, only 6.5 percent came this close. [30]

A Role for the EB-66

After Rolling Thunder ended in 1968, EB-66's began making increasing use of chaff, a trend that continued during the 1972 fighting over the North. When dispensing chaff, the planes avoided SAM-infested areas and relied on the wind to carry the radar reflectors over the target. The EB-66 orbits, based on up-to-date weather forecasts, covered an area measuring 4 by 40 nautical miles, although the plane might fly either an elliptical or a figure-eight flight path. On board each aircraft were two hoppers, each carrying 348 chaff bundles. The EB-66 might drop its first bundle 3 hours or more before the strike it was helping to screen, depending on wind velocity and direction and distance to the target. These factors also determined the rate at which chaff was dropped, usually 12 bundles per minute from each of the two hoppers. [31]

In 1972, the EB-66 fleet was old and tired yet rose again for its last battle.
Source: U.S. Air Force.

As they had earlier, the EB-66s once again tried to jam radio communication between ground controllers and MiG interceptors. Results varied according to the distance the jamming signal had to travel and the angle formed by the orbiting EB-66, the radio transmitter, and the interceptor formation. The shorter the distance and narrower the angle, the better the results. [32]

The planes also continued to engage in long distance electronic jamming of radar. On most missions, they focused upon target acquisition, early warning, and ground control intercept sets, though EB-66Es sometimes jammed Fan Song transmitters located near the perimeter of the area protected by SAMs. On a Fan Song mission, an EB-66C usually came along to watch for signals from gun-laying radars, such as Fire Can, and alert the E model which could then direct a noise barrage against them. Jamming acquisition and early warning radars also paid indirect dividends, forcing Fan Song and gunlaying sets to pick up the approaching aircraft, thus presenting targets for anti-radiation missiles. [33]

The EB-66's were most effective against radars guarding targets on the perimeter of North Vietnam. But their jamming signal grew weak at long range, and they rarely ventured within range of SAM concentrations. "Stand-off jamming, " as a JCS study has pointed out, "cannot adequately degrade terminal threats along the penetrating aircraft's flight path unless the radar is forced to look in the direction of the jamming platform." Stand-off jamming nevertheless remained the standard EB-66 practice, for the improved results obtained by closing with the target seldom outweighed the greater risk involved. [34]

Lt. Col. Iceal E. Hambleton was a USAF ballistic missile expert with a Top Secret/SCI clearance and his capture by the North Vietnamese Army would have been of tremendous benefit to them and the Soviet Union. His rescue was the subject of the film BAT-21. Source: U.S. Air Force.

A dramatic demonstration of EB-66 vulnerability took place on 2 April 1972 over an area just south of the demilitarized zone, where the presence of SAMs was suspected but not yet confirmed. An EB-66C, flown by Maj. W. L. Bolte, and an EB-66E were supporting a three-plane B-52 strike. The enemy launched four SAMs, all of which passed harmlessly between the bombers and the EB-66s. Five minutes later, a battery located almost 20 nautical miles south of the demilitarized zone fired three missiles, probably aiming at the jamming source. One of these hit Bolte's plane at an altitude of 24,000 feet and sent it shrieking earthward, flames streaming from both wings. At 18,000 feet, the left wing broke off, and seconds later the plane crashed. As the plane plummeted to its destruction, the navigator,

Lt. Col. Iceal E. Hambleton, parachuted safely. For 11 days, he matched wits and endurance with North Vietnamese infantrymen, until rescued by friendly troops near Dong Ha, South Vietnam. This loss resulted in a decision to ban EB-66Cs, which generated less jamming power than the E models, from areas where SAMs were suspected. [35]

Occasionally, the risk to the EB-66s seemed justified. It was a case of vulnerability be damned, on 18 December, the first night of Linebacker II, when a flight of three E models assumed an orbit just 40 nautical miles west of Hanoi. While these planes attempted to blanket the radar-controlled defenses of North Vietnam's capital, a flight of MiG-21s closed to within 5 nautical miles, cutting off retreat to the west, but patrolling Phantom jets intervened and chased the enemy away. Missile batteries then opened fire, forcing the EB-66Es to dodge a half-dozen SAMs and temporarily disrupting their jamming pattern. [36]

The age-related problems with the EB-66 continued to escalate. The 1972 campaign was the swan-song for the old warrior; when redeployed home after the war, the type was withdrawn from service. Source: U.S. Air Force.

Despite the enemy's reaction, these planes survived. Indeed, the only EB-66 lost during Linebacker II was an E model that crashed on 23 December while attempting to land at Korat. The accident. which was attributed to engine failure, killed Maj. George F. Sasscer. the pilot, and his two electronic warfare officers, Maj. Henry J. Repeta and Capt. William R. Baldwin. [37]

Mechanical problems persisted throughout 1972, though usually with less tragic results, so that the 42d Tactical Electronic Warfare Squadron, with an average of 19 EB-66s, was hard pressed to fly the 15 sorties per day required of it. In September, a shortage of replacement engines caused the number of daily sorties to decline to eight. Even as spare engines were arriving in Thailand, leaking fuel tanks reduced the daily average of operationally ready aircraft for October to just six, one third of the number on hand. The arrival of a sealant for the tanks restored the readiness rate to about 50 percent for the rest of the year. [38]

Two other problems that surfaced during Rolling Thunder -- interference with friendly Navy electronics and stereotyped action -- recurred in 1972. EB-66 interference with the Navy's floating control center proved easy to resolve. When the difficulty surfaced, the EB-66 crews simply avoided jamming that would overlap frequencies used by Navy controllers. [39]

The problem of rigid orthodoxy of EB-66 tactics was particularly evident in the support of B-52 strikes against the panhandle of North Vietnam. The number and model of EB-66 used, the presence of fighter cover and Wild Weasel aircraft and the dispatch of drogue-equipped tankers combined to disclose where the Stratofortresses would attack. For instance, when Wild Weasels and an F-4 escort teamed with one EB-66, the target lay between the demilitarized zone and the 18th parallel. If the B-52s were headed farther north, their counter-measures support consisted of two EB-66s escorted by four F-4s. The refueling orbit of the drogue-equipped tankers indicated whether the EB-66s were screening an attack on North Vietnam or Laos. An alert enemy, a security survey concluded, could assemble all this information from radio call signs, with his agents on or near American air bases providing verification. [40]

Changes in Wild Weasel Tactics

Wild Weasel tactics during 1972 differed from those employed earlier. The radar suppression flight, which now reached the vicinity of the target well before the strike force, assumed responsibility for protecting the recently initiated chaff flights. The hunter-killer team reappeared, with F-4Es as killers and F-105Gs or newly arrived Wild Weasel versions of the F-4C as hunters. Despite these changes, however, the basic goal of Wild Weasel remained radar suppression, and the two Rolling Thunder anti-radiation missiles, the AGM-45 Shrike and the AGM-78 Standard ARM, continued in use.

The closely interrelated revisions of Wild Weasel tactics reflected changes made since 1968 in fighter-bomber operations. Instead of maintaining the 1-minute interval adopted during Rolling Thunder, for instance, Wild Weasel flights again were entering the target area as much as 20 minutes ahead of the strike force, timing formerly rejected as needlessly dangerous. The new schedule was necessary because radar suppression aircraft had to protect the chaff flight, which it was especially vulnerable to enemy missiles because of the radar-reflecting corridor forming behind it. Also, since the increased time interval again exposed the Wild Weasels to MiG attack, they received fighter cover unless F-4s and tankers were simply not available. [41]

The teaming of F-4Es with Wild Weasel F-105Gs, or occasionally F-4Cs, in hunter-killer flights became necessary because anti-radiation missiles no longer ensured radar suppression. After November 1968, Wild Weasels stopped carrying bombs, relying exclusively on Shrike and Standard ARM to deter enemy operators. These tactics worked well enough during the lull ending in the spring of 1972, for F-105Gs were dealing with isolated radars on the fringe of the area protected by SAMs. During Freedom Train and Linebacker, however, the Iron Hand flights led the way into heavily defended regions where they encountered veteran radar men adept at detecting the launch of an AGM-45 or AGM-78 and capable of reacting immediately. Their reaction typically consisted of shifting to passive tracking in order to foil the Shrike, while another similar radar began transmitting from a different quadrant, thus confusing the Standard ARM memory circuit. Once the Wild Weasels had expended the two missiles each normally carried, the radar operators could transmit with impunity. [42]

F-4C Wild Weasel flying over North Vietnam, December 1972. Unlike the F-105G, the F-4C Wild Weasel could not carry Standard missiles. Source: U.S. Air Force

In order to maintain suppression, the Wild Weasels resumed carrying bombs, a practice that further hindered the performance of the F-105Gs, already laden with countermeasures equipment and anti-radiation missiles, and sometimes the bulky Standard ARM. A better solution was the revival of the hunter-killer idea, using a F-105G, without bombs, to find targets for a bomb-laden F-4E, which could carry more ordnance than the Wild Weasel. The Phantoms had still another advantage; in the event of MiG attack, they could jettison their bombs, wheel about from the rear of the maneuver pod formation, and engage the enemy. Consequently, beginning in the summer of 1972, these hunter-killer teams shared the radar suppression mission with Iron Hand flights made up exclusively of Wild Weasels. [43]

Despite the reintroduction of the hunter-killer team radar suppression remained the primary goal; the destruction of a transmitter was an infrequent bonus. The Wild Weasels might achieve suppression merely by launching an anti-radiation missile at a known or suspected radar, preventing it from coming on the air, but the limited number of missiles carried by the aircraft meant that the threat of bombs or gunfire was necessary to enforce the deterrent. A Wild Weasel tactical manual declared that "freefall high explosive ordnance comprises the optimum killing capability in the Weasel arsenal." This threat was obvious to radar operators, who realized that once they revealed their location, no electronic trickery could deflect the standard aerial bomb. Strafing was a less desirable weapon against radar sites, however, for the attacking aircraft had to dive within range of light antiaircraft guns and automatic weapons. As a result, the 20-mm aerial cannon was a "munition of last resort, " to be used with "great prudence." [44]

On a typical mission led by a hunter-killer team, the F-105Gs might carry Shrikes that homed on either Fan Song or Fire Can and Standard ARMs for use against ground control intercept and height finder radars. These Wild Weasels tried to suppress the target radars by feinting toward known transmitters. Sometimes the aircraft actually launched missiles timed to home on the set when it attempted to track the strike force; at other times they merely showed their readiness to launch. While the strike force dropped its bombs, the hunter-killer team usually split into two elements, with the F-4Es bombing any SAM or antiaircraft battery that gave away its location by opening fire. Usually the four-plane team reunited to cover the withdrawal of the fighter-bombers, though the Phantoms might remain behind to finish attacking the weapon sites. When the main force had departed, the team hit previously located gun and missile positions. The strength of the defenses and the number of fighter-bombers challenging them determined how many hunter-killer or Wild Weasel flights engaged in radar suppression. [45]

Hunter killer group of F-105G Wild Weasels and F-4Es take fuel on the way to North Vietnam for a LINEBACKER strike in the summer of 1972. Source: U.S. Air Force

During the 1972 fighting, problems with the Standard ARM reduced Wild Weasel effectiveness. As early as mid-April, the missiles, along with the adapters used in carrying them, were being expended faster than they could be replaced. Accidental loss of empty adapters, however, came to an end after a simple change in F-105G wiring enabled the pilot to jettison the fuel tank under the left wing without simultaneously releasing the AGM-78 adapter attached under the opposite wing. Shipments of Standard ARMs from bases elsewhere built up stocks in Southeast Asia, but production of the B model, the only type being used in combat, had stopped, ensuring that the missile shortage would occur. When Linebacker II ended the Wild Weasels had only 15 of the weapons on hand.

Besides being in short supply, the Standard ARM was beset with malfunctions. The failures, which amounted for a time to 30 percent of the weapons launched, occurred when the rocket motor failed to ignite, the homing device did not lock onto the target, or the warhead exploded a few seconds after launch. The only feasible solution was to substitute Shrikes for Standard ARMs. [46]

Another hazard, excessive radio traffic, caused by several pilots calling out warnings after sighting the same SAM contributed to the loss of at least one Wild Weasel. The plane went down north of Hanoi on 29 September because repetitive SAM alerts choked the radio channels and prevented the Weasel crew from hearing the warning that might have saved the aircraft. By this time, radio had become essential to Wild Weasel survival. Since the warning gear on board these aircraft often was inundated by radar signals, the human eye was more reliable than electronics in detecting missiles, and whoever saw a SAM depended on radio to pass on the information.

The best way for Weasel crews to foil the SAM was to see the missile and outmaneuver it. Jamming equipment was of limited value because the signal could interfere with the homing device in both the Shrike and Standard ARM. Consequently the planes carrying these weapons refrained from jamming unless actually fired upon. Because Wild Weasel airmen had to see missiles to be sure of eluding them, low-lying clouds presented a mortal threat. Linebacker experience demonstrated that the planes needed a safety margin of 8000 feet above the cloud deck in order to react to SAM launchings. [47]

As was the case during Rolling Thunder, the contribution of Wild Weasel to 1972 operations was difficult to estimate. Between 1 April and 30 September, Air Force crews launched 678 Shrikes and 230 Standard ARMs, a total of 908 anti-radiation missiles. These scored just 3 confirmed and 96 possible hits. Yet 320 of the weapons were launched to prevent the enemy from coming on the air, rather than to destroy or silence a radar already transmitting. A careful analysis of intelligence and operational reports disclosed that anti-radiation missiles had forced 185 active fire control radars to shut down. In some cases, Fan Song transmitters had fallen silent while guiding SAMs toward a target. Navy and Marine Corps planes that used these same missiles reported similar results: 1,425 launches, 254 of them preemptive, resulted in 33 confirmed and 38 possible hits plus 521 instances when enemy radar abruptly shut down. [48]

Dangerous Latecomer: The SA-7

Following the North Vietnamese invasion of the South on 30 March 1972, the enemy introduced a new antiaircraft missile, the lightweight SA-7. Fired from the shoulder like the World War II bazooka, the SA-7 consisted of a smooth-bore fiberglass tube with optical sights, an infrared homing rocket, a trigger mechanism, and a small thermal battery for power. Both launcher and battery were expendable, but the trigger mechanism could be detached and reused.

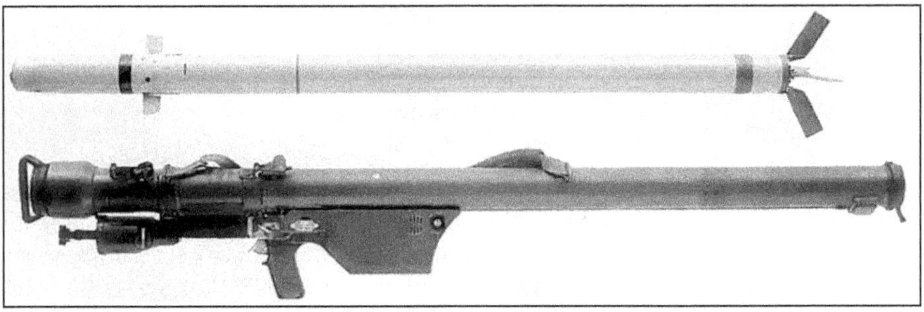

The SA-7 Igla missile posed a new threat to tactical aircraft. It made obsolete an entire generation of slow, light-strike aircraft . Source: U.S. Air Force

The launcher measured 146.9 centimeters (58.75 inches) with sight and trigger attached. The missile was 140 centimeters (56 inches) long, 68.8 millimeters (2.75 inches) in diameter, and weighed 9.1 kilograms (20.3 pounds). When fired, after the missile cleared the tube, two sets of fins popped into position. One set, located aft of the rocket booster nozzle, consisted of four fins arranged to ensure stability by imparting a counterclockwise roll during night. The other set, two fins mounted about 27.5 centimeters (11 inches) from the nose, were the control surfaces. The SA-7 could be lethal against aircraft that were as far as 2 nautical miles from the launcher and at altitudes up to 10,000 feet. The explosive charge, the equivalent of 0.565 kilograms (1.3 pounds) of TNT, either detonated on impact or destroyed itself 20 seconds after launch.

The gunner had to see the target before he could engage it with the SA-7. Consequently, after switching on the thermal battery, he located the aircraft in the optical sight and then waited for a warning tone, which together with a light in the aiming device, told him he was locked onto the target. The gunner then pressed the trigger halfway to uncage the gyroscope within the missile and continued pressing until the weapon fired. The projectile homed on the heat source provided by the target's engines and exhaust.

Because electronic countermeasures did not affect the SA-7's uncooled lead sulfide detector, other techniques had to be used. These included installation of thermal shielding on helicopters, the release of flares as decoys, and outmaneuvering or outrunning the airborne missile. Experience soon showed that aircraft flying above 9000 feet and at speeds faster than 500 knots were generally safe. Also, rapid changes of heading and altitude could vary the intensity of the heat detected and confuse the homing device. The jet pilot who saw an SA-7 could evade it by making a sharp, fast turn, but if he used his afterburner to gain speed, he provided a better target for the heat seeker.

Flares quickly became an essential item of equipment for aircraft that would be required to fly low and slow over contested areas. Source: U.S. Navy

Helicopter crews, forward air controllers, and other "slow movers" were in the greatest danger from SA-7s. Pilots of piston-powered or turboprop aircraft found that a steep bank into the missile could help shield the exhaust and engine heat upon which the weapon was homing. Helicopters, moreover, could combine this maneuver with an unpowered windmilling descent.

Between the March 19 72 invasion and the January 1973 truce, American and South Vietnamese aircrews reported sighting 528 SA-7s. The heat-seeking missiles destroyed 45 aircraft and damaged 6 others. A late addition to North Vietnam's air defenses, the weapon saw little action north of the demilitarized zone, for it was used primarily for battlefield defense rather than for protection of airfields, railway yards, and similar installations. [49]

Linebacker II: The First Two Nights.

By the time the B-52s launched Linebacker II, in December 1972, Fan Song operators had polished their skill in combating the bombers. As early as 9 April, near Dong Ha in South Vietnam, a SAM damaged an attacking B-52, the first missile hit suffered by a Stratofortress in seven years of Southeast Asia operation. On that day the first warning came when an electronic warfare officer on board one of the bombers in the three-plane cell detected a guidance signal. After the second B-52 in the cell had dropped its bombs and begun turning away, at least three missiles bored toward it. One of these exploded just 50 feet from the plane's left wing tip, puncturing the external fuel tank on that side and tearing into the fuselage. Despite the damage, Capt. Kenneth J. Curry landed safely at Da Nang. The absence of a Fan Song signal prior to launch indicated that the enemy had tracked the jamming source. [50]

When the Stratofortresses went North, SAM defenses were ready. The first of these planes damaged by a missile over North Vietnam was a B-52D hit by fragments while attacking Thanh Hoa on 23 April. Not until 22 November, however, did one of the SAMs destroy a B-52, a Thailand-based plane bombing a target 24 nautical miles northwest of Vinh. The pilot, Capt. Norbert J. Ostrozny, tried to fly the crippled aircraft to Nakon Phanom but lost control about 12 nautical miles from that base. The entire crew parachuted and was rescued within a few hours. [51]

In shooting down Ostrozny's B-52D, North Vietnamese radar had provided accurate guidance despite electronic jamming by the victim, the other two bombers in the cell, and three EB-66s orbiting some distance away. As he approached the target, Captain Ostrozny had received additional protection from a chaff corridor created by four F-4s and from two Iron Hand F-105Gs. Throughout the approach, when the countermeasures were most effective, the radar operator had contented himself with passive tracking, simply following the jamming source across his scope. He allowed the B-52 to soar unchallenged to the release point. Then, as the plane turned sharply away after dropping its bombs, the wings formed an angle of roughly 45 degrees with the horizon, and the strongest part of the jamming cone passed ineffectually beyond the SAM site. At this instant, the Fan Song transmitted just long enough to pinpoint its target before the missile battery launched the two SAMs that exploded beneath the B-52. [52]

SAM controllers thus demonstrated that they had learned to take advantage of certain technical and tactical characteristics of the B-52. When Linebacker II began, the enemy already knew that the plane's jamming transmitters were least effective during sharp turns, that each Stratofortress habitually made just such a turn after dropping its bombs, and that formations usually attacked from about 35,000 feet. Armed with this information, the Fan Song operator passively tracked the jamming signal to determine azimuth and elevation, used the normal operating altitude to establish the range, then verified the range by transmitting for a couple of seconds as the three-plane cell was making its post-target turn.

These tactics minimized or entirely avoided exposure to anti-radiation missiles, and they enabled the enemy to catch the B-52 when it was most vulnerable to radar controlled weapons. For the defenders, however, waiting for the post-target turn had an obvious failing, it permitted the bomber to reach the release point opposed only by barrage fire. [53]

A B-52 being shot down over Hanoi. The hit on the aircraft is actually the smaller explosion right at the top of the picture. The huge ball of fire in the middle of the frame is the fuel from the aircraft's tanks falling and then exploding. Such fuel cascades often burn from the bottom up, giving the appearance of another missile. Finally the picture also has two lens reflections to the right and left of the main event. All these effects can make photo interpretation difficult and are the root of many conspiracy theories. Source: Vietnamese People's Liberation Army

The 207 B-52s in Southeast Asia carried an impressive array of countermeasures equipment that included the ALT-22 jamming transmitter, used against the Fan Song track-while-scan beacon, and the ALT-28, which could either reinforce the ALT-22 or engage in down-link jamming. By mid-December, all B-52s serving in this area mounted four ALT-28s and three ALT-22s, except for 41 of the 98 G models on Guam that carried older, less powerful ALT-6Bs instead of the ALT-22s. Electronic warfare officers in the bombers usually directed two ALT-28s and two ALT-22s or ALT-6s against Fan Song and used the other pair of ALT-28s to jam the SAM guidance beacon. With the remaining ALT-22 or ALT-6, he usually attacked height-finder radars. [54]

Plans for Linebacker II B-52 strikes drew upon the experience gained during the spring of 1972. Out of respect for North Vietnamese antiaircraft guns, which had proved so deadly against Rolling Thunder fighter-bombers, the B-52s flew beyond reach of even the 85-mm weapon. At these altitudes, SAMs were their most dangerous, but the bombers maintained the three-plane cells which thus far had provided adequate countermeasures protection. To avoid mid-air collision, the cells kept 2 minutes apart.

During the first three nights, the Stratofortresses sought to minimize exposure to the SAMs by getting out of missile range as rapidly as possible, even though steeply banked turns of 113 to 160 degrees were required. The officers who approved these tactics realized that such a turn was "a characteristically vulnerable position" because the "effects of both TWS and beacon jamming were minimized.

"[55] They believed, however, that the greater speed in leaving the target area would more than offset the loss of jamming coverage.[56]

Linebacker II began on 18 December with 121 B-52s in three waves attacking seven targets near Hanoi. Supporting the first wave were 19 countermeasures aircraft: three EB-66s for standoff jamming, eight chaff-dispensing F-4s, and eight Wild Weasels for radar suppression. The second wave received the same countermeasures support, but as the third wave arrived, four Navy A-7's replaced the eight Wild Weasels, and five EA-3Bs joined the EB-66s, their Air Force counterparts, in long-range jamming.[57]

Everything went well as the waves approached their targets. Capt. D. D. McCrabbe, an electronic warfare officer on board one of the bombers, felt that the SAM crews were "a little confused" at first, but the confusion ended as soon as the raiders had dropped their bombs. "We started doing our post-target turn and just all hell broke loose, " the captain related. "They just started throwing everything at us." The enemy fired 164 missiles that night, downing three aircraft and damaging two others. All the hits occurred as the B-52s were turning into wind of at least 71 knots and struggling to get out of SAM range.[58]

Wreckage of "Rose-1", a B-52D shot down over Hanoi. Source: Vietnamese People's Liberation Army.

Pending completion of a detailed evaluation by the Security Service and the Air Staff, planners made minor adjustments in countermeasures support. For radar suppression, the first wave on the 19th could count on eight F-105Gs, the second had ten and the third had four Navy A-7s. Eight F-4s laid chaff corridors for the first wave, two Phantoms sowed additional chaff for the second, and eight Phantoms preceded the final wave. Once again, five Navy EA-3Bs joined three EB-66s already in covering the third wave.[59]

The three B-52s lost the first night had been flying at 34,000 feet, 38,000 feet, and 38,500 feet, and indication that the lowest and highest aircraft in the bomber streams had not received adequate chaff protection. As a result, base altitudes of 34,500 and 35,000 feet were established for the bomber cells, keeping all the aircraft closer to the center of a corridor sown from 36,000 feet. And, in order to give the B-52s room to avoid approaching SAMs, the interval between cells was doubled to 4 minutes. This increased interval allowed the enemy more time for passive radar tracking, but the increased chaff coverage, plus the jamming barrage from the bomber cells, were expected to offset this advantage. [60]

No really drastic revision of B-52 tactics seemed necessary. As one of the aircraft commanders later pointed out, "there really was no reason to change tactics at this point." The Stratofortresses, after all, had used the same basic procedures over Haiphong on 15 April and suffered no losses. "Sure," he admitted, "we lost some aircraft on the first day [of Linebacker II], but the area we flew in was better defended." On the second night, 19 December, no B-52s were lost, though two suffered damage, which seemed to confirm these views. [61]

Hard Lessons are Learned

The tactics that worked on the second night failed on the third, as the sky came alive with 221 SAMs, 39 more than had been fired on the 19th. "Just watching all the SAMs," said Capt. Bruce Kordenbrock, "was like watching a show until you realized they were starting to shoot at you." To Capt. D. W. Jameison, the enemy seemed to be "just salvoing off like six SAMs at a time," of which "maybe one or two would be tracking." Other B-52 crewmen, such as Maj. L. M. Sweet, reported hectic radar activity on the night of the 20th, with as many as three Fan Song radars simultaneously tracking a single plane. A tail gunner saw three missiles, obviously guided from the ground, pursue his bomber through a hard left turn and explode within 750 feet of him. [62]

Below the bomber stream, Captain Kordenbrock saw a stratum of antiaircraft shells bursting so close together that "you could get out and walk on it." The scene reminded him of "all the war movies you've ever heard about." TSgt C. M. O'Quinn, the tail gunner whose B-52 had been chased by three missiles, believed "they were sending up triple A [antiaircraft artillery] and SAMs together, hoping we'd dive to avoid the SAMs and fly through the flak." Although the B-52s remained above 34,500 feet and escaped flak damage, missiles downed six bombers and damaged a seventh. [63]

Until the Security Service could evaluate the role of countermeasures on the night of the 20th, three interim actions were taken to reduce SAM effectiveness during the strikes planned for the 21st. First, electronic warfare officers shifted one jammer from the Fan Song track-while-scan beam to reinforce the downlink jamming barrage. Second, planners decided to compress the bomber stream, stacking the cells from 33,500 to 38,000 feet and timing their arrival over the targets at intervals between 90 and 120 seconds. This decision reversed the 19

December policy of extending the bomber streams to keep the aircraft close to the center of the chaff corridor. Finally, plans for the night of 21 December called for two of the bomber streams, 24 of the 30 attacking aircraft, to avoid sharp turns, relying on their speed, boosted by the prevailing wind, to approach from the west and to depart eastward over the Tonkin Gulf. [64]

Another memorial to the air battle over Hanoi in December 1972. This pile of wreckage is claimed to be a B-52 but even cursory examination shows this not to be so. The centerpiece appears to be a Navy A-6 while the rest can best be described as "miscellaneous junk". There is even an old piston engine in there. Source: Vietnamese People's Liberation Army.

The countermeasures evaluation for the first night, which became available on the 23d, confirmed that the three bombers lost to SAMs on 18 December had received fatal hits during post-target turns, when their jamming was least effective. More significant was the disclosure that all three had belonged to "the cells in which no jamming was committed to the beacon (downlink) frequency." [65] The shift of emphasis from Fan Song to downlink jamming therefore appeared wise.

Initial comments on chaff effectiveness on the night of 18 December implied that the corridors had been satisfactory, despite winds that varied from predictions by as much as 10 degrees in direction and 14 knots in velocity. Subsequent reports stated, however, that the three Stratofortresses shot down that night had been on the fringe of a ragged corridor. The difference between the predicted winds and the air currents actually encountered had disrupted the planned chaff coverage. [66]

When the Security Service analysis of the 20 December strikes arrived, it disclosed that chaff protection had been non-existent that night. Plans had called for 27 of the 33 attacking cells to be protected by chaff corridors while within range of SAM sites. Although "many cells flew through some portion of the corridor" during the night's operation, "only four cells were actually in chaff at their respective bomb release lines and post target turns." The six B-52s that fell victim to SAMs were from 5 to 10 nautical miles from the chaff concentrations when hit. The wind had again blown gaps in the coverage, forcing planners to think in terms of widespread blankets rather than comparatively narrow corridors. A shortage of chaff, however, caused postponement of this modification of counter-measures tactics. [67]

The evaluation of the third night's countermeasures effectiveness also sustained the decision, made in time for the 21 December operation, to substitute speed for steeply banked turns in getting out of SAM range. The Security Service reported that the missile batteries had again concentrated their fire upon B-52s in the post-target turn. In one case, the intended victim foiled the enemy by quickly leveling off and "reinjecting the self-protection noise jamming element during the terminal phase of the intercept." [68]

As was entirely predictable, the North Vietnamese placed their anti-aircraft weapons and fighters in civilian areas to gain immunity from attack. Source: Vietnamese People's Liberation Army

Study of the first three nights also led to a prohibition against using certain B-52 models over Hanoi-Haiphong. One of the three B-52s shot down on the 18th, and four of the six destroyed on the 20th were B-52Gs carrying ALT-6 jamming transmitters instead of the newer ALT-22. For the rest of Linebacker II, therefore, bombers mounting the ALT-6 were restricted to less heavily defended targets. [69]

As the struggle against the SAM continued additional Navy aircraft joined the support force, with EA-6Bs in engaging in stand-off jamming and A-7Es flying SAM suppression. Effective 24 December, chaff blankets replaced the corridors, and the bombers began releasing chaff in self defense. The B-52 electronic warfare officers received instructions to drop the radar reflectors whenever they detected Fan Song tracking signals during any turn sharper than 45 degrees and with a bank angle of 30 degrees or more. Although designed to confuse airborne radar, the chaff would also provide some extra protection during the dangerous post-target turn. [70]

Throughout Linebacker II, countermeasure tactics underwent analysis and revision. Tests at Eglin AFB indicated that the B-52 antenna radiation pattern was ill-suited for downlink jamming. In addition, the enemy seemed to be using a modified SAM, fitted with a more powerful guidance beacon that was less susceptible to a modulated noise barrage. As a result, the B-52s that attacked North Vietnam on 26 December employed only two ALT-28s against the downlink. All other transmitters jammed the track-while-scan beam, except for one ALT-22 or ALT-6 directed against height finder radars and the I-band T-8209 signal. When the electronic warfare officer detected the T-8209, he was to switch on his ALR-18 jamming transmitter. Although designed for the airborne radar carried by MiG interceptors, this device also had some value against the T-8209. [71]

There were rumors and reported sightings of SA-3 missiles during Linebacker II but these were never confirmed. Russian sources suggest that the missile wasn't supplied to Vietnam during the war in case it was compromised by the Chinese. Source: Vietnamese People's Liberation Army,

The attacks of 26 December incorporated all the countermeasures lessons learned since the 18th. Plans for the operation combined precise timing with countermeasures protection, sending 120 bombers to attack 10 targets within 15 minutes. Careful selection of approach and departure routes brought the B-52s over seven targets almost simultaneously. Also, altitudes and interval varied to confuse the defenders, with some cells climbing or descending during the shallow post-target turn. Counter-measures support required 62 Air Force and Navy planes for stand-off jamming, SAM suppression, and chaff dispensing. [72]

The 24 F-4s laying chaff on 26 December put down two blankets, one west of Hanoi and the other over Haiphong. The six B-52 cells attacking Thai Nguyen, some 30 nautical miles north of Hanoi, had no chaff cloud to protect them, but certain of the bombers released chaff during the post-target turn, whether or not a Fan Song signal was detected. [73]

Despite the resourcefulness of those who planned the raid, SAMs claimed two B-52Ds on the night of 26 December. scoring both hits in the vicinity of Hanoi. One aircraft received the fatal damage before releasing its bombs, but the other went down during its post-turn. The countermeasures plan was not at fault, however, for in both cases, one aircraft in the cell had turned back, depriving the remaining pair of vital jamming power. [74]

MiG-21s scored no B-52 kills. Source: Vietnamese People's Liberation Air Force.

Linebacker II Recapitulation

In the 11 nights of Linebacker II, SAMs destroyed a total of 15 B-52s and damaged 9. Following the loss of six planes on 20 December, the third night, missile effectiveness declined rapidly. In eight nights from the 21st through the end of the operation on the 29th (no bombing took place on the 25th). SAMs downed six B-52s, the same number that had perished on the 20th, and damaged four, one fewer than had sustained damage during the first three nights.

As Linebacker II progressed, the enemy's SAM defense grew weaker. On 18 December, 164 of the missiles soared into the night sky. The defenders of Hanoi and Haiphong fired 182 the following night and 221 on the third, but the number of launchings plummeted to just four missiles on 23 December, rebounded to 73 on the 27th, then dropped to 23 on the final night of the bombing. Intelligence confirmed the launching of 567 SAMs in the first three nights but only 315 during the remaining eight. Obviously, the North Vietnamese were running out of missiles. By mining the harbors, bombing the rail and road net, and attacking SAM support facilities) American airmen had prevented the enemy from stockpiling enough SAMs to replace those expended during the first three nights. 75

Linebacker II SAM Firings [76]			
Date	Number SAMs	B-52s Lost	% Hit Rate
18	164	2 Gs, 1D	1.83%
19	182	0	0.00%
20	221	4 Gs, 2Ds	2.71%
21	40	2 Ds	5.00%
22	43	0	0.00%
23	4	0	0.00%
24	16	0	0.00%
25	0	0	N/A
26	68	2 Ds	2.94%
27	73	2 Ds	2.73%
28	48	0	0.00%
29	23	0	0.00%
TOTAL	**882**	**6 Gs, 9Ds**	**1.70%**

In terms of effort, Linebacker II represented the most intensive use of electronic countermeasures of the Vietnam war. The losses suffered on the first three nights resulted in tactical adjustments that enhanced the effectiveness of countermeasures support. Once again, the attackers had met the challenge and overcome a skillful defense, helped on this occasion by a deepening shortage of SAMs.

VII. CONCLUSIONS

Early in the air war against North Vietnam, the Air Force embarked upon Project Corona Harvest, an attempt to document its wartime experience in order to derive lessons from the Southeast Asia fighting. The Corona Harvest final report, which appeared in the summer of 1973, touched upon the subject of electronic countermeasures. Those Air Force officers who drafted this document called attention to the improvement in the defenses of the North and the consequent refinement of countermeasures equipment and techniques.

From what the Corona Harvest report called "an air defense capability which, for modern times, was considered as basically similar to that faced in World War II and Korea, " the radar-controlled defenses of North Vietnam rapidly evolved into what this same summary described as "the most formidable electromagnetic threat ever encountered by U.S. tactical forces." [1] American electronic countermeasures generally kept pace, enabling the air offensive to continue. According to Corona Harvest, the "limited EW capability" possessed by the Air Force in the spring of 1965 improved throughout the Vietnam fighting. A key to this improvement in equipment and technique was the continuing analysis of countermeasures effectiveness by the Air Force Security Service. [2]

In the summer of 1966, while attempting to answer a Department of Defense inquiries about countermeasures effectiveness, Air Force headquarters had difficulty finding reliable data upon which to base a reply. As a consequence, the Chief of Staff, Gen. John P. McConnell called upon the Air Force Security Service to devise some method of determining the tactical impact of electronic warfare. The Security Service set up a small working group within the Air Force Special Communications Center at Kelly AFB. Texas. and in March 1967 systematic reporting and analysis began. Besides disseminating "flash reports" prepared immediately after a day's strikes, the group used computers to recreate for detailed study those missions or incidents of special interest to combat units. The analysts also prepared periodic summaries showing trends in electronic warfare. [3]

Air Force sources, however, did not provide adequate electronic intelligence needed for a comprehensive assessment. To fill the void, the Air Force turned to the National Security Agency (NSA), which agreed to furnish information from its radar and communications intercepts. The NSA material offered an insight into the reaction of North Vietnam's air defenses to countermeasures and to other aspects of electronic warfare. [4]

Conclusions

Air Force, Navy, and Marine Corps intelligence, with the help of NSA, provided adequate information for the prompt evaluation of countermeasures activity. In disseminating the evaluations, however, the Air Force Security Service ran into difficulty. Because of the sensitive nature of some of the intelligence, and the risk of compromise in forwarding it to operational units, tactical commanders were not informed of the basis for the conclusions contained in the electronic warfare evaluations. [5]

Despite this problem, the system of evaluation and reporting enabled the Air Force to make the best possible use of its countermeasures activity. The equipment did not always work miracles, however. Some of it was old, and even the newer items imposed limitations on the performance of the aircraft that used them.

Wild Weasel enjoyed the best reputation of any the countermeasures employed in Southeast Asia. The Security Service credited the aircraft's anti-radiation missiles with a kill probability of just 5 percent -- 1 of 20 missiles launched could be expected to score a destructive hit -- but conventional munitions carried by the planes proved even more destructive. Corona Harvest concluded that Wild Weasel had "effectively carried out" the suppression of "radar directed guns and missiles," [6] and Security Service analysts agreed, declaring that Wild Weasel despite the small number of hits with anti-radiation missiles, had limited the enemy's use of radar, thus reducing the number of missiles he could launch with accuracy and causing a decline in the likelihood of SAM kills. [7]

Under other circumstances, the Wild Weasels ht have tried to destroy rather than to suppress. During the Vietnam war, however, too few aircraft were available to press sustained attacks against SAM defenses. In addition, the Shrike missile had a short range and no memory circuit, while the less numerous Standard ARM was frequently unreliable. [8]

Although the first QRC-160 self-protection pods used in Southeast Asia failed miserably, the improved QRC-160A-1 and later types became lifesavers for Air Force fighter-bomber crews. These devices jammed SAM radars, enabling the attacking formations to fly at altitudes where the missiles could be deadliest and thus remain above the flak concentrations that had proved so dangerous early in the war. The self-protection pod deserved the tribute paid by Col. Robin Olds, who late in 1967 hailed it as "the most significant development in the air war over North Vietnam." [9]

The Air Force officers who wrote the Corona Harvest final report agreed that whenever fighter-bombers went after heavily defended targets, self-protection pods "were effective in degrading SAMs and radar directed AAA. " The devices had certain drawbacks, however. As this same group pointed out, pods "displaced ordnance, prevented optimum antenna placement, and required a particular type of formation flying that did not always provide the best protection from MIG attack." [10]

Unlike the pods, which Corona Harvest considered the most valuable where defenses were strongest, the old and vulnerable EB- 66 countermeasures aircraft "were forced from strike escort to a stand-off jamming role, resulting in reduced ECM effectiveness." Despite the panel's dismissal of this elderly plane, the final report acknowledged that the Air Force would continue to need a large, extensively equipped jamming craft until it acquired a self-protection pod capable of protecting individual fighter-bombers against all types of radar. A vulnerable aircraft like the EB-66 was inadequate to the task of reinforcing and supplementing the barrage laid down by current types of pods. A new airplane was needed, the Corona Harvest report declared, and "this airplane should be capable of operating in a high threat environment with a degree of survivability equal to the strike force it supports." [11]

Chaff came into its own during Linebacker II despite a discouraging start. According to radar scope photography, high winds at first played hob with the corridors, scattering the chaff, and stripping the protection from 80 percent of the B-52 cells carrying out attacks from 18 through 24 December. However, chaff blankets employed from 26 through 29 December almost reversed these statistics, with 84 percent of another night's bomber force receiving the planned coverage. Not a single B-52 was downed or damaged while within a dense chaff concentration, though the Security Service acknowledged that other countermeasures undoubtedly contributed to this immunity. [12]

The radar warning equipment available to fighter-bomber crews in Southeast Asia had two major weaknesses. The receiver was vulnerable to friendly jamming, so that the pilot of a pod-equipped aircraft had to shut down from time to time in order to detect hostile radar. In addition, the warning gear did not always indicate the specific weapon site threatening the aircraft. Over North Vietnam, the number and variety of radars concentrated in so small an area frequently saturated the warning device, forcing the pilot to rely on his eyesight to spot SAMs fired at him. But, in spite of these failings, the equipment did furnish a general warning that told aircrews to commence jamming and begin looking for missiles ascending toward them. [13]

At first glance, the losses sustained early in Linebacker II might seem to confirm previous doubts that B-52 countermeasures were adequate for missions over the Red River delta. The enemy, after all, had succeeded in burning through the noise barrage, and the possibility existed that the initial use of ALT-6s in conjunction with ALT-22s might have actually highlighted the bombers for Fan Song operators. A closer examination indicates, however, that overall tactics, might have countermeasures failure, were at fault. Losses might have been fewer if the B-52s had tried from the outset to saturate the defenses, approaching from different directions, attacking several targets simultaneously, and avoiding the steeply banked turns that diverted a plane's jamming pattern. [14]

The Corona Harvest report revealed that stereotyped strike tactics were a problem for most of the war. Reliance on daylight fighter-bomber attacks, usually delivered at about the same hour against targets concentrated in a compact area, gave the defense a marked advantage. Electronic countermeasures came to take the place of surprise against an alert and resourceful enemy. But jamming also tended to follow familiar patterns and at times may even have told the enemy what sort of mission to expect. Clues provided by countermeasures procedures could, for example, have alerted the North Vietnamese to certain drone reconnaissance flights and helped them predict some B-52 targets. [15]

According to Corona Harvest, one of the most important countermeasures lessons learned from the Vietnam war was that better preparation was needed for any future conflict. The attacking Americans and the North Vietnam defenders had waged a seesaw contest for electronic supremacy, with the offense finally winning. In other circumstances, however, the Air Force might not be able to duplicate the spectacular progress that characterized the Vietnam struggle. During 1966, for example, both the Shrike-carrying Wild Weasel and the QRC-160A-1 self-protection pod emerged as dependable countermeasures. In the future, however, highly trained American airmen should be ready to commit thoroughly modern equipment upon the outbreak of war, use the gear effectively, and accurately evaluate its impact on the enemy.[16]

APPENDIX: AIRBORNE CONTROL OF FIGHTERS

The four F-105D Thunderchiefs of Zinc flight were circling over North Vietnam on 4 April 1965, waiting their turn to pounce on Thanh Hoa bridge, an important Rolling Thunder target. The pilots could hear radio chatter about MiG fighters in the area, and the leader of another flight warned everyone "to keep their heads up." Seconds later, the pilot of Zinc 03 caught sight of "four aircraft coming in from behind us out of a 20 degree dive approximately 3,000 - 4,000 feet behind the flight. " He radioed a warning and with his wing man, Zinc 04, broke toward the attackers and jettisoned the two external fuel tanks and eight 338-kilogram (750-pound) bombs that burdened each of the Thunderchiefs.

Their abrupt turns made the two planes lose speed and prevented them from intervening as the grayish-colored fighters -- now easily recognizable as MiG-17s -- flashed past and opened fire on the lead element of Zinc flight. Within seconds, the flight leader and his wing man were coaxing their mortally damaged Thunderchiefs toward the Gulf of Tonkin where they parachuted to safety. [1]

Things the EC-121D Couldn't Do

For several weeks before the destruction of the two F-105Ds, various headquarters had been discussing the deployment to Southeast Asia of an EC-121D task force to provide MiG warning for Rolling Thunder formations. Actually, the idea of using the Lockheed-built airborne radar platforms had originated the previous summer, when planners realized that the carrier strikes of August 1964, in response to the Tonkin Gulf incident, might possibly trigger North Vietnamese air raids against the South. The use of EC-121Ds for airborne early warning did not seem especially urgent, however, until February 1965 when raids against the North began on a systematic basis. In March, with Rolling Thunder underway, representatives of Headquarters, Air Defense Command, and of the 552d Airborne Early Warning and Control Wing, which would provide the necessary planes, journeyed to Hawaii to work out with Pacific Air Forces the details of a JCS-directed deployment. [2]

Plans formulated as a result of this meeting called for the EC-121D detachment to extend the coverage of the existing radar network which, at the time, consisted of stations at Monkey Mountain, near Da Nang, South Vietnam, and at Nakhon Phanom in Thailand. The Monkey Mountain radar could track aircraft at almost any altitude throughout most of South Vietnam, while operators at Nakhon Phanom covered North Vietnam south of Hanoi and the adjacent portion of the Tonkin Gulf. The EC-121Ds would keep watch over the Red River delta and the region to the north. Besides providing early warning of air attack against the South, airborne radar would warn of "any enemy activity directed toward U.S. strike aircraft. " Other likely assignments included coordinating search and rescue activity, giving emergency navigation assistance, and helping fighter-bombers low on fuel to rendezvous with aerial tankers. Thanks to this planning, the Big Eye Task Force (Officially, Detachment 1, 552d Airborne Early Warning and Control Wing) of 5 EC-121D's stood ready to depart from McClellan AFB, California, when deployment orders arrived on 4 April 1965. [3]

The EC-121D Warning Star. This aircraft is the famous "Triple Nickel" that, on Oct. 24, 1967, over the Gulf of Tonkin, guided a U.S. fighter into position to destroy a MiG-21. This action marked the first time a weapons controller aboard an airborne radar aircraft had ever directed a successful attack on an enemy plane. Source: U.S. Air Force.

Ironically, the destruction of the two F-105Ds, which prompted the Big Eye deployment, later evoked a reluctance to send the EC-121Ds northward over the Tonkin Gulf. Again and again the same question arose at 2d Air Division, the headquarters that exercised operational control over the task force: Could an EC-121D survive if it ventured north of the 17th parallel? Unaffected by these doubts, the task force staff insisted that a small escort would be enough, just a few fighters under control of the EC-121D itself.

A few weeks after arriving at Tan Son Nhut air base near Saigon, members of the Big Eye contingent began wondering if 2d Air Division had any intent on of using their skills. A bewildering series of test missions sapped morale, as the task force attempted a variety of duties for which it was ill equipped. The EC-121Ds quickly demonstrated that they lacked the communication gear required to direct air strikes in support of ground troops and that their small and inconveniently located windows ruled them out as aerial observation posts for B-52 attacks. Some officers, the task force history reports, thought these seemingly bizarre assignments "as an exercise to prove how many things an EC-121 couldn't do." [4]

These suspicions proved unfounded. When dispatched on missions for which it was suited, the EC-121D justified the confidence of those who flew and maintained it. Yet the lack of enthusiasm within air division headquarters was understandable, for this was an old airplane that took up valuable space at already crowded Tan Son Nhut and carried electronic gear unfamiliar to most Air Force pilots.

Airplanes, Electronics, and Men

Nicknamed the Connie by its crews, the Lockheed EC-121D was a military adaptation of the 4-engine piston-powered Super Constellation commercial transport, but dorsal and ventral housings for radar antennas now marred the graceful silhouette of the airline version. Located atop the fuselage was the antenna for the height finder radar while beneath the plane was the antenna for the search radar. The height finder, though credited with a maximum range of 120 nautical miles, seldom picked up an aircraft much beyond 70 nautical miles. The search radar had a nominal range, under ideal conditions, of 250 nautical miles, but this varied according to the altitude of the EC-121D and the nature of its target, and seldom exceeded 180 nautical miles. [5]

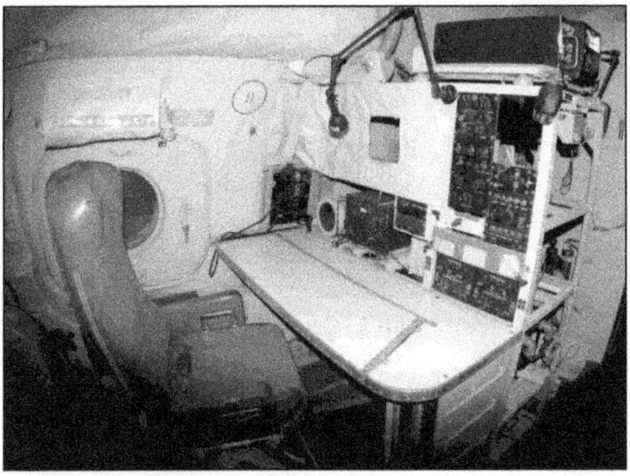

Radar operator's station on an EC-121D. Source: U.S. Air Force.

The crew of a Big Eye aircraft usually numbered 18. Up front was the flight crew, consisting of pilot and copilot, two navigators, two flight engineers, and a radio operator. The mission crew was responsible for all the electronic equipment located amidships in the operations center, included the senior director, a duty controller, a crew chief and his assistant, four radar operators, an intercept control technician, and two radar maintenance specialists. The senior director and duty controller were responsible for the instructions issued by Big Eye to friendly aircraft. [6]

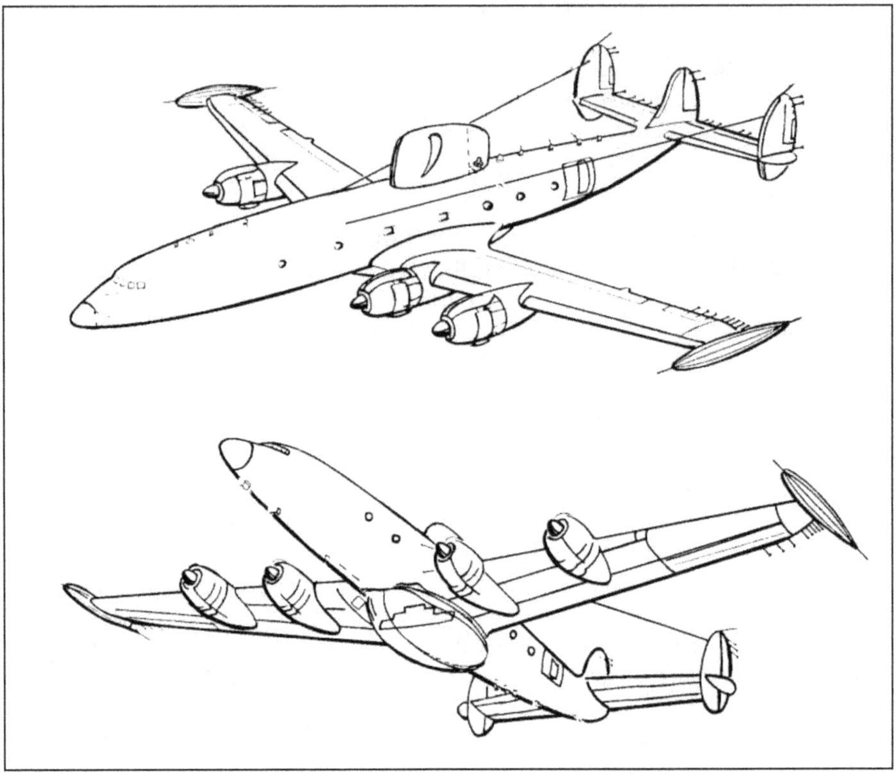

Although the EC-121D had been designed primarily to detect hostile bombers flying high above the ocean, it proved effective in maintaining short-range radar surveillance of low-flying reconnaissance planes taking photographs over Cuba during the 1962 missile crisis. North Vietnam, however, posed a greater challenge than Cuba, for the EC-121Ds would have to track a larger number of planes at ranges in excess of 150 nautical miles, operating in a similarly confined area. These aircraft, moreover, would be flying over a comparatively large land mass, and the Connie's search radar became confused by reflections from terrain features and objects on the ground. Fortunately, Big Eye crews were able to apply a lesson learned off Cuba and maintain altitudes as low as 50 feet so that the search radar beam was reflected from the surface of the Tonkin Gulf, thus raising the line of sight and avoiding ground clutter. [7]

By using this search radar technique, operators obtained the general location and heading of hostile aircraft. Since most radar contacts took place beyond range of the height finder, Big Eye could not determine the altitude of North Vietnamese MiGs. Since EC-121D did not yet have the equipment for precise control [8] the best the task force could do was alert the covering fighters, whose crews then had to rely on eyesight or airborne radar to close with the enemy.

Although limitations in the search and height finder radars handicapped the task force's airborne controllers, the Connies carried an electronic device that helped to compensate for these failings. This article was the APX-49 recognition set, which was capable of identifying friendly aircraft at ranges as great as 200 nautical miles. This equipment transmitted a coded signal which triggered the transponder carried by these planes. The transponder reacted by emitting a distinctive signal that was picked up on radar scopes within the EC-121D. Once the operator had decoded the reply, he knew for certain which aircraft had responded.

The recognition set was not infallible, however. To make the system work friendly pilots had to cooperate by switching on their identification gear so that the transponder would react. Often an operator had to make two or more sweeps before receiving a transponder signal clear enough for him to decode. Moreover, decoding was a manual operation, a serious drawback in view of the amount of traffic under surveillance. [9]

The communication gear carried by the first EC-121Ds to see service in Southeast Asia lacked the necessary range and reliability. The three ARC-27 ultra high frequency radios installed in the planes were already obsolete, though the aircraft's pair of more powerful ARC-85s proved generally satisfactory despite maintenance problems.

Radio operator's station. Source: U.S. Air Force.

To help compensate for Big Eye's deficiencies, a SAC KC-135 radio relay plane was tested over Southeast Asia in September 1966. This variant of the four-jet Boeing aerial tanker automatically relayed radio messages either to American tactical aircraft or to the control centers that directed them. The EC-121Ds, however, did not receive a really modern radio until the introduction of the ARC-109 in September 1968. [10]

At first, Big Eye radio operators had to encode and decode all messages containing security information, a time consuming procedure since most operational traffic fell into this category. Not until December 1966 did they obtain the use of a "scrambler" that ensured security of voice transmissions. [11]

The EC-121D, known at the time as the RC-121C, first entered Air Force service in 1954. Although a decade had passed between the acceptance of the first of these planes and the Vietnam deployment, the aircraft proved reliable and durable during their Southeast Asia tours. Structural corrosion, the most serious maintenance problem encountered there, was an inevitable result of sustained operation in the tropics. [12]

The EC-121D in flight. Many of the early missions on the Alpha orbit were at MUCH lower altitude than this. It was a great relief when the Connies could return to higher altitudes. Source: U.S. Air Force.

Big Eye's first radar surveillance missions tested both the courage and endurance of those on board. A task force history tells of pilots skimming the Gulf of Tonkin and peering through tropical showers to detect the masts of fishing vessels hidden by the downpour. This was a more dangerous and demanding kind of flying than high-altitude surveillance where the automatic pilot did much of the work.

Nor was the Connie a comfortable aircraft. The air conditioning system had been designed for commercial transport flying at medium altitude, not for a military airplane jammed with heat-producing electronics and flying at wavetop height over a tropical sea.

All things considered it was, as the task force history said about these early flights, "a credit to their self discipline" that officers and men "performed under this pressure without incident." An occasional glimpse of China's Hainan Island, as the plane executed its orbit, served to remind cockpit crew of the gloomy warnings from 2d Air Division headquarters that they could not survive north of the 17th parallel. To the equipment operators enclosed within the fuselage, any abrupt change of direction summoned up the nightmare of collision with a mast and impact against the water. When MiG attacks failed to materialize, however, and changes in equipment permitted higher altitude operation, the pressure relaxed. [13]

Big Eye's Evolving Mission

On 16 April 1965, a pair of EC-121Ds took off from Tan Son Nhut airfield to head out over the Gulf of Tonkin. One entered a racetrack orbit roughly 90 nautical miles southeast of Hanoi. Flying from 50 to 300 feet above the water, it used the reflected beam of its search radar to track aircraft over the North and employed recognition gear to distinguish which were friendly. The second plane backed up the first from a medium altitude station farther from shore. Alpha track, the orbit nearer the coast, provided search radar coverage of all aircraft higher than 8000 feet above North Vietnam. From the other, Bravo track, EC-121D crewmen used recognition gear to monitor flights over the North and radar to cover the Gulf of Tonkin. [14]

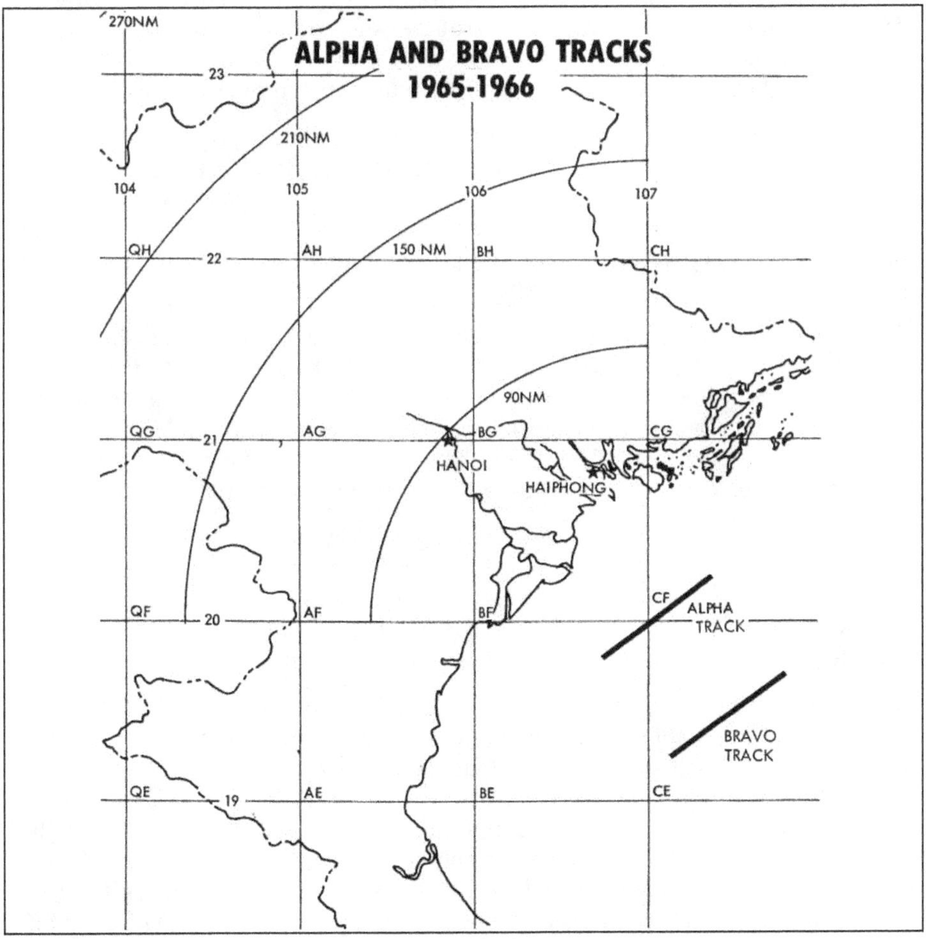

From these two orbits, the Big Eye task force issued MiG warnings "in the blind," alerting the entire fighter cover rather than any specific flight and locating the enemy according to his distance from a common reference point, the city of Hanoi. Until automatic relay equipment appeared, these general warnings were retransmitted by controllers on the ground, a time-consuming procedure when dealing with supersonic fighters. The Phantoms nearest the MiGs responded, searched them out, and attacked. [15]

Imprecise though it was, this kind of warning sometimes gave the covering F-4s a tactical advantage that proved fatal to North Vietnamese airmen. On 10 July 1965, Capt. William P. Reboll Jr., the senior director in an EC-121D flying 50 feet above the Tonkin Gulf, issued a MiG warning that resulted in the destruction of two hostile planes. An F-4C manned by Captains Kenneth E. Holcombe, and Arthur C. Clark shot down one of the MiG-17s, and Captains Thomas S. Roberts and Ronald Anderson, also in a Phantom, received credit for downing the other. These were the first Air Force victories of the war, and the attendant excitement obscured the fact that they were due more to chance -- the Phantoms being in the right place when the alert came -- than to the information provided by Big Eye. [16]

Tracking Air Force planes was not especially difficult, provided the pilots cooperated. As Capt. Jerry Kaffka, a member of the Big Eye Task Force, recalled, "We worked out a system whereby we can identify a specific flight of ours" by assigning an identification response to one plane in the flight. When the operator interrogated this signal, the plane's transponder automatically replied, giving the position of the flight. [17]

According to Kaffka, the presence of Navy carrier planes over the North complicated for a time the work of the Big Eye Task Force. Early in the war, Naval aviators did not use identification equipment except when approaching the aircraft carriers, a practice which caused some uneasy moments on board the EC-121Ds. Navy planes would "go in, say down in the panhandle and come up over land, and then be coming back out toward us at about 600 knots at 30,000 feet."

With eight or ten different tracks approaching on radar, the airborne director could only summon the fighter cover, sometimes as few as two F-4Cs or F-104s, and "expend a heck of a lot of time in just running identification passes to see who they were. " The Navy's willingness to adjust to circumstances and have its pilots "squawk" -- use their identification gear -- solved this problem. [18]

While monitoring air traffic in order to give MiG warning, Big Eye airborne directors also were able to help pilots low on fuel to rendezvous with KC-135 tankers. SAC, which provided the Stratotankers used in Southeast Asia, at first insisted that the vulnerable KC-135s stay south of the 19th parallel, and all the EC-121Ds could do was advise the fighter pilot of the most direct route to the nearest orbiting tanker. The men serving in the KC-135s soon realized how helpful the task force could be and frequently defied policy by crossing the 19th parallel to meet fighter-bombers critically low on fuel. Upon receiving word from a pilot in trouble, the Big Eye radioman alerted the nearest tanker and requested it to start north. Although not obliged to respond, the SAC crew usually complied, and F-105 and F-4 pilots soon began addressing emergency requests directly to the Big Eye aircraft, asking assistance in arranging a refueling rendezvous. This improvisation worked well, but all parties had to be circumspect in reporting their activity.

SAC's official attitude changed early in 1967. One factor causing this change was a test, conducted over Laos, in which an EC-121D successfully coordinated an aerial refueling. Another was the success of the informal arrangement between Big Eye senior directors and tanker commanders. One of the directors, Maj. Theron M. Perry, earned the Distinguished Flying Cross by arranging a rendezvous between a KC-135 and a fighter-bomber so low on fuel that its jet engine flamed out as it locked onto the tanker's boom. The decisive factor, however, was the insistence by General Momyer, the Seventh Air Force commander, that the EC-121Ds control the tankers during refueling emergencies. [19]

Navy EC-121s shared the airborne surveillance mission. Source: U.S. Navy

Another activity in which the task force became involved was search and rescue. The use of EC-121Ds in that role was inevitable, since their equipment and location enabled them to maintain surveillance over areas not covered by ground stations or shipboard radars. Some of the more hectic action, however, took place almost within sight of American warships. In January 1967 for instance, Capt. Gerald D. Long, an airborne controller, released the two F-104s serving as Big Eye fighter cover so they could escort to Da Nang a KC-135 damaged in a mid-air collision. Scarcely had the Starfighters returned when one of them suffered a control failure, forcing the pilot to parachute into the gulf. Captain Long immediately contacted a nearby radar control ship, which launched a helicopter that rescued the pilot just minutes after he had bailed out. [20]

Occasionally the task force assumed the defensive role envisioned for it when the question of deployment first arose in 1964. For example, American intelligence specialists at Saigon became concerned during October 1965 that the North Vietnamese might attack Da Nang with their Ilyushin IL-28 twin-jet light bombers. Until the new Dong Ha radar site, some 25 nautical miles south of the demilitarized zone, was activated, low-flying IL-28's could approach Da Nang undetected. To prevent this, Big Eye on 5 October established a nighttime orbit over the gulf that permitted coverage of the area north of Da Nang. No additional planes or crews were available to assume the increased burden, so consequent crew fatigue and postponed aircraft maintenance threatened Rolling Thunder support. Fortunately, construction at Dong Ha proceeded on schedule, and Big Eye was able to abandon air defense surveillance on 19 November. [21]

Meanwhile, the task force had become involved in issuing border warnings to friendly aircraft about to trespass over the People's Republic of China. The first such intrusion occurred when the EC-121Ds were on the ground at Tan Son Nhut. Without Big Eye to follow his progress, Maj. Philip E. Smith, USAF, the pilot of a Starfighter, strayed over Hainan Island, was shot down, and remained a prisoner of the Chinese until released in March 1973. The incident caused the Joint Chiefs of Staff to remind the 2d Air Division of the importance of making full use of the Connies in tracking friendly aircraft. [22]

Shortly afterward, the Chinese protested a second border violation. The 2d Air Division staff summoned the Big Eye operations officer, the senior director on the low-level mission flown the day the alleged incursion took place, with all logs and charts for that flight. Luckily, the senior director had been alert to the possibility of a border violation in attacking one of the day's Rolling Thunder targets and had issued instructions to make overlays showing the tracks of all flights -- American, North Vietnamese, and Chinese -- that ventured near the border. These overlays supported the contention that Rolling Thunder aircraft had respected the international boundary, showing instead that Chinese planes had made a brief incursion over North Vietnam. Ushered into an inner sanctum at air division headquarters, the senior director repeated this information over a scrambler telephone circuit to the National Military Command Center at the Pentagon, where the Joint Chiefs of Staff were investigating the incident.

Border warning promptly became another of Big Eye's missions, with the surveillance normally performed from the higher altitude, Bravo, orbit, using recognition equipment. Whenever a squawking flight penetrated a buffer zone, established with varying depth along the Chinese border, a director on board the EC-121D radioed a code word, usually changed each day, to warn the pilot away. [23]

The border warning was unpopular among fighter pilots. Although they readily admitted they might become disoriented while dodging SAMs or fighting MiGs and stray over China, few could see any point in respecting the border when the consequences could be death or imprisonment in North Vietnam. Some aviators went so far as to shut down their identification gear when flying near the border in order to retain freedom of action and avoid a possible reprimand should they enter

Chinese air space. Others did not turn on their transponders because they believed the North Vietnamese were using recognition equipment patterned after the American gear to trigger a signal and thus pinpoint victims for MiGs or SAMs. [24]

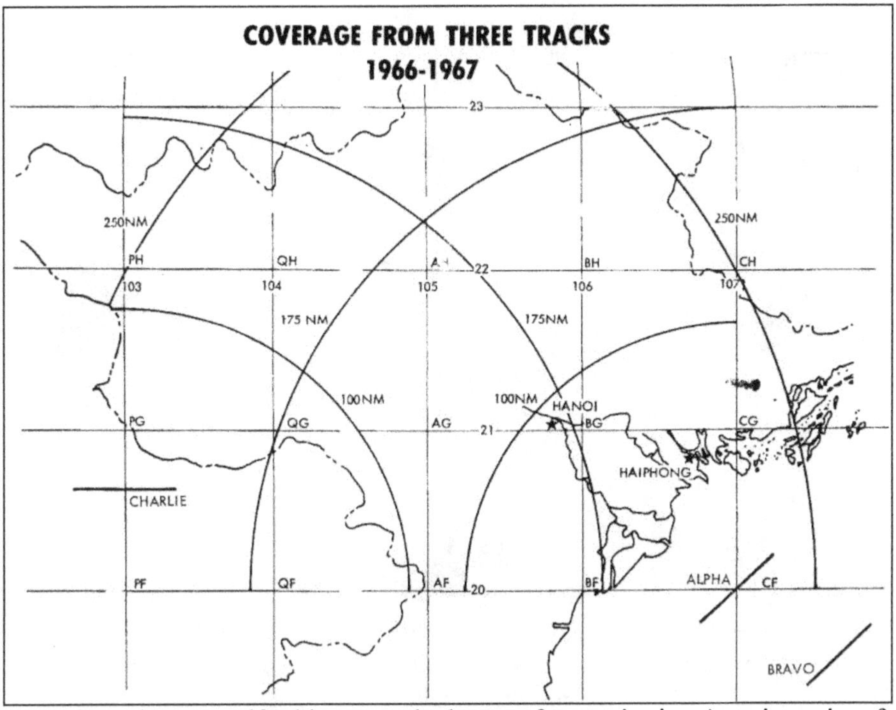

For a time, the two gulf orbits seemed adequate for monitoring American aircraft swarming over the North. Then, on 12 May 19 66, an F-105D chased a North Vietnamese MiG-17 some 25 nautical miles inside China and shot it down. A board of inquiry headed by Brig. Gen. Robert G. Owens, Jnr, USMC, discovered that the low-altitude Alpha orbit had not been flown that day because the assigned EC-121D suffered an equipment failure. The Owens Board concluded, however, that a plane on Alpha station could not have covered the area where the incursion took place. The board therefore recommended establishing an additional orbit over Laos, north of the 19th parallel. [25]

On 24 June, an EC-121D conducted a test to determine what coverage could be obtained from a point just north of Vientiane. Beginning in July, the test aircraft maintained an altitude of 10,000 to 15,000 feet while orbiting over the 20th parallel. Results proved successful, and by the end of October, EC-121Ds were flying this inland orbit, called Charlie station, whenever Rolling Thunder strikes were scheduled. Lacking planes and crews enough to cover Charlie, Alpha, and Bravo simultaneously, the task force juggled its resources, either shifting a plane to Charlie from Bravo or canceling the Laos mission when weather did not permit strikes near the Chinese border. When reinforcements arrived in May 1967, the

College Eye contingent -- redesignated from Big Eye on 9 March of that year -- was able to fly all three orbits until December, when improved monitoring equipment, specifically the QRC-248 interrogator, made Alpha unnecessary.[26]

EC-121s over the Gulf of Tonkin usually had a pair of F-104s or F-4s as escorts. Source: U.S. Air Force.

The Owens Board also gave impetus to the establishment of a Tactical Air Control Center, North Sector, near Da Nang. On 1 November 1966, therefore, a battle commander at the center began monitoring Air Force operations over North Vietnam, passing along MiG and border warnings and information on rescue activity, frequently merely retransmitting information from Big Eye.[27]

From its inception, the Tactical Air Control Center, North Sector, had the potential to become the hub of an automated control network of radio relay and radar surveillance aircraft. Modernized EC-121s would be equipped to trigger either friendly or enemy identification transponders, and with communication sets to transmit these data, in coded form, to the battle commander. These projected EC-121 modifications ran afoul of rising costs and unforeseen delays, so that the end product, the EC-121T, did not join the College Eye Task Force until 1971. The EC-121Ds did undergo various interim improvements, however, which increased their effectiveness in handling friendly fighters. The installation of this equipment enabled task force controllers to direct F-4 Phantoms against MiG interceptors instead of merely issuing a warning that the enemy was approaching. [28]

Equipping for More Active Role

One far-reaching improvement, the addition of the QRC-248 interrogator, was a result of continuing surveillance of Cuba. During March and April 1965, as Rolling Thunder was gathering momentum in Southeast Asia, a prototype QRC-248, being tested at the Air Defense Command's Key West, Florida, radar station, successfully interrogated the Russian-built transponder on a Cuban aircraft at a range of about 200 nautical miles. The prospective advantages of such a device in the skies over North Vietnam were so exciting that Air Force headquarters quickly approved the installation of one of the few available sets in an EC-121D for tests, first in the United States, then by a 30-day combat evaluation in Southeast Asia.

Initially the new interrogation device proved less spectacular in combat than it had in the Florida test. In order to avoid alerting the North Vietnamese to this new form of surveillance, QRC-248 operators refrained from actually triggering enemy transponders, instead limiting their role to passive reception of signals from identification devices queried by North Vietnamese controllers. Because of this restriction, early results were disappointing. [29]

As the mission portfolio of the EC-121s expanded, so did the number of bumps, lumps and other excrescences on their once-sleek fuselages. Source: U.S. Navy

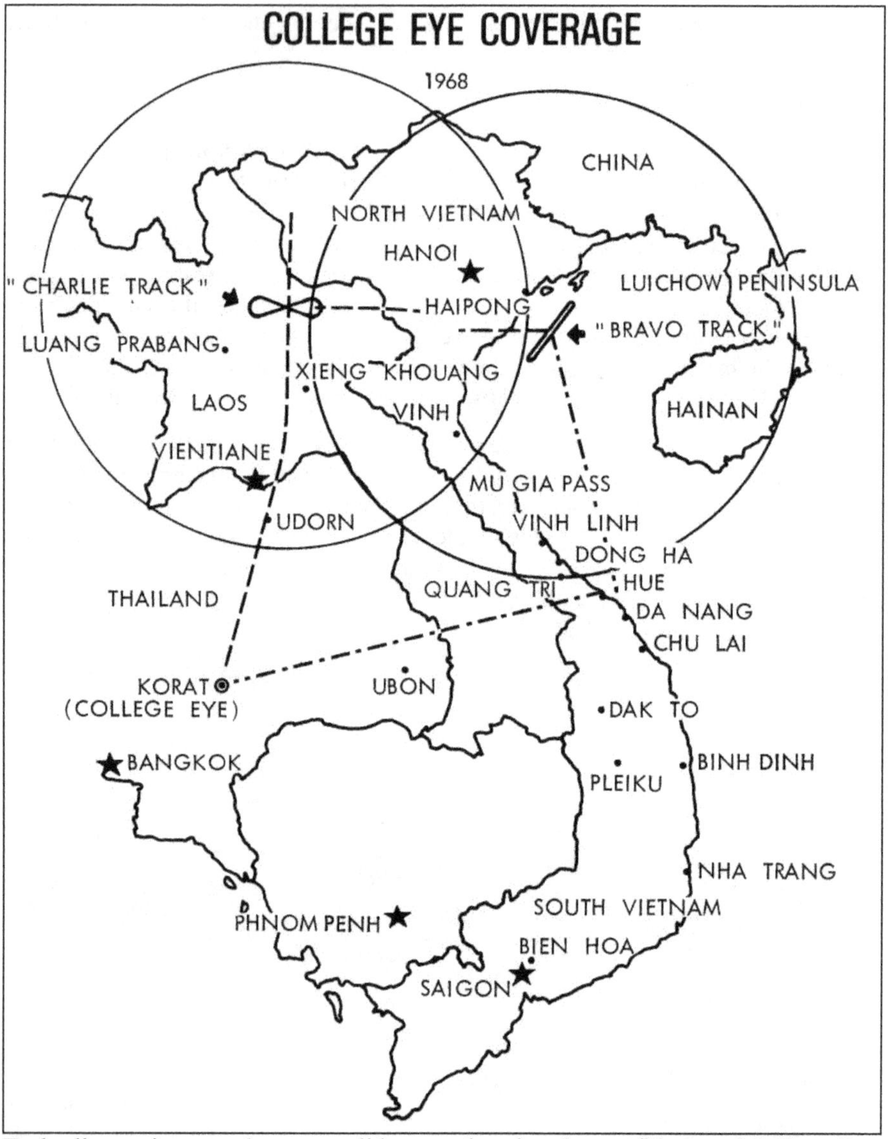

Early disappointment, however, did not undermine the confidence of College Eye pilots in the potential of the QRC-248, and some pilots of the 8th Tactical Fighter Wing, who shared this confidence. Together the airmen worked out informal procedures whereby airborne directors passed information on MiG activity directly to F-4 crews instead of routing it through the North Sector control mechanism. Formal tests followed, and in October 1967 General Momyer's head-quarters approved the arrangement, authorizing College Eye to provide the combat air patrol with range and vector to the enemy instead of the more general MIG warning issued before. [30]

To sharpen coordination between College Eye directors and F-4C crews, pilots from the 8th Tactical Fighter Wing not only received an orientation but went along as observers on actual EC-121D missions. On 23 October, Majors Joseph Moore and William Kirk flew from Ubon, where the task force had been based, to Udorn for this type of familiarization. Major Moore joined the crew of the plane selected for the next day's Bravo mission -- an EC-121D called Triple Nickel because of the three 5's on its tail number. Major Kirk returned to Ubon to fly combat air patrol for a strike scheduled for the afternoon of the 24th. The day got off to an auspicious start for Triple Nickel when the senior director, Capt. Joseph McGrath, picked up several distinct returns during the morning, dispatched F-4Cs to intercept, and contributed to the possible destruction of a North Vietnamese MiG. For a time, however the mission rested in the hands of a radar technician, TSgt James Bleeker. "Heat, " he recalled, "caused a resistor to burn out on our main scope," but he "tied the resistor together with string and put the scope back in action."

In the afternoon, Major Kirk arrived on station. As Major Moore watched, Captain McGrath acquired a target and radioed a vector to Kirk's flight, which intercepted the enemy. Closing in from behind, Kirk launched a heat-seeking missile that exploded near the tail of a MiG-21. A series of rapid turns ensued before Kirk was again in position to attack. He then launched two more missiles, one of which inflicted further damage, and finally finished off the plane with 20-mm fire. The North Vietnamese pilot parachuted as his MiG plunged to earth. [31]

Increasing reliance on the QRC-248 led to cancellation of the low-altitude Alpha track. From station Charlie over Laos and from Bravo, 10,000 feet above the gulf, two EC-121Ds fitted with this equipment could monitor North Vietnam's heartland and the border with China. By the end of November, the task force commander was reporting that the low-altitude Alpha station had become "an expensive luxury," and before the end of December it was abandoned. [32]

Even with the QRC-248, fighter control continued to be a demanding job. One College Eye director, Capt. Richard Carle, discovered this when he found himself in a situation where the airborne automatic radio relay failed him, a flight of MiGs without transponders appeared on radar, and several of the F-105Ds in the area shut off their identification gear. As the action unfolded, one segment of the F-105D strike force saw MiGs approaching, jettisoned its bombs, and turned away from the target, heading in a direction from which another Thunderchief formation was approaching. Simultaneously, several of the escorting F-4s attempted to head off the MiGs. Four groups of aircraft -- the F-4s, the MiGs, and the two F-105 contingents -- were now converging at velocities approaching the speed of sound. Captain Carle got through the F-4s, warned them that "multiple friendlies" were dead ahead and thus averted a calamity. Both the Americans and North

Vietnamese emerged unscathed from the melee that erupted when the Phantom crews and MiG pilots caught sight of each other. [33]

Airborne controllers like Captain Carle utilized QRC-248 in conjunction with the recognition gear used to identify friendly aircraft. Unfortunately, the latter could not cope with the number of aircraft simultaneously in action over the North. Work had therefore begun on an automatic recognition and decoding device, the GPA-122. which was successfully demonstrated at Van Nuys, California, in November 1966. Development of this new equipment for spotting friendly planes was sluggish until October of the following year, when MSgt Gary Walker rounded up the necessary component parts, assembled one of the sets, and tested it in an EC-121D at McClellan AFB where it performed as expected. Nevertheless, despite Walker's "quick look" test, it was June 1968 before College Eye received enough of the devices to equip all its planes. [34]

A Potential Rival

Late in July 1967, another Connie arrived in South-east Asia. The newcomer was Rivet Top, an EC-121K, a prototype which the Tactical Air Warfare Center was testing for use as an airborne command post equipped to direct the complex aerial war being waged over the North. The electronic array installed in Rivet Top was designed to provide the airborne commander with up-to-the-minute data on every aspect of the enemy's defenses. [35]

The EC-121K. Source: U.S. Navy

To deal with one component of North Vietnam's air defenses, the SAM, Rivet Top boasted radar direction finding equipment plus a computer that contained data on as many as 100 launch sites. The computer compared every radar signal received with the information stored in its memory and automatically revised its data bank, as necessary. Armed with this intelligence, the airborne commander could issue missile warnings and direct attacks on sites where the radar transmissions originated.

Rivet Top was designed to steal enemy radar returns, enabling the airborne operator to photograph the image being seen by an enemy radarman on the ground below. Since the radar return disclosed principal terrain features like mountains or lakes, intelligence analysts could compare this picture with existing aerial photos to locate the transmitter. [36]

These feats were beyond the electronic ability of College Eye EC-121Ds, but Rivet Top, in addition to these, could also control fighters, a job the task force claimed as its own. Indeed, the EC-121K carried control equipment not found in the Ds. Seated before special consoles in Rivet Top's crowded interior, communication specialists eavesdropped on radio traffic between enemy ground controllers and interceptor pilots. Moreover, in addition to the College Eye recognition gear, the K model boasted equipment capable of interrogating two Russian-designed transponders that were immune to the proddings of QRC-248. One was the SRO-1, used in some early MiGs and based on an identification set furnished the Soviet Union by its World War II allies. The other was the SOD-57, originally intended as an aid to airport traffic controllers but used over North Vietnam to help ground control intercept operators determine the heading and altitude of their aircraft. [37]

Operations analysts at Seventh Air Force headquarters compared the performance of College Eye, Rivet Top, and the Navy's EC-121M, Big Look, which also could trigger identification devices that did not respond to QRC-248. These comparisons placed College Eye's EC-121Ds dead last in ability to detect and report hostile aircraft, a finding that seemed to jeopardize the task force's future. [38]

The officers assigned to the College Eye Task Force rallied to the defense of their organization. According to Col. Ross Davidson, task force commander, the analysts had based their conclusion on the erroneous premise that more equipment meant more MiGs detected. This was not so, he argued, for discussions with both fighter pilots and his own controllers had convinced him that the QRC-248 was responsible for 90 percent of the information obtained on MiG traffic. Since all three EC-121 variants used this same piece of equipment, any difference in effectiveness had to stem from operating technique. Here College Eye was at a disadvantage. Task force operators, still laboring under rules designed to prevent compromise of the QRC-248, employed a rapid sweep and focused a minimum of energy on possible targets. Colonel Davidson was confident, however, that his EC-121Ds could do as well as Rivet Top or the EC-121M, if the operators were allowed to reduce antenna speed and spotlight the enemy. [39]

The views of the task force commander apparently were accepted, for College Eye and its EC-121Ds remained on the job, sharing the fighter direction mission with Rivet Top and then taking over after that aircraft returned to the United States in September 1968. The improved equipment that Rivet Top had used to identify hostile planes was tested by College Eye in 1968 and became standard in the EC-121T, which replaced the D models beginning in 1971. [40]

The EC-121D did undergo major modification as result of Rivet Top's success in controlling fighters. This was the only role in which Rivet Top was an unqualified success. The plane's radar locating equipment however, became confused in the maze of signals emanating from North Vietnam and did not live up to expectation. Ling-Temco-Vought corporation was awarded a contract to add communications intercept equipment to 14 Aerospace Defense Command planes. The first of these so-called Rivet Gym aircraft flew a test mission over the Tonkin Gulf on 5 May 1 68, some five weeks after President Johnson suspended the bombing north of the 20th parallel. The newly installed equipment generated enervating heat that forced the exhausted operators to shut down so delicate electronic components could cool. Despite this problem, results were judged satisfactory, and in two weeks the College Eye task force was using the original Rivet Gym alternately with TAC's Rivet Top. During September 1968, when Rivet Top departed for the United States, Ling-Temco-Vought was finishing work on the last Rivet Gym aircraft. [41]

According to a College Eye veteran, the communications intelligence obtained by Rivet Gym could be "classed as 'nice to have," but was not "absolutely essential to a defensive warning system." This officer, Capt. Richard M. Williams, was quick to add, however, that Rivet Gym did provide "extremely valuable information for offensive anti-MiG employment, such as heading, altitude, intent, etc." [42]

A Band of Brothers

During 1967, College Eye finally found at Korat a satisfactory base, the fourth it had occupied since deploying to Southeast Asia in April 1965. While flying out of Tan Son Nhut, the first of these airfields, task force members complained that they were treated like poor relations. Offices were crowded, and the operations section did not have enough safes to store the flood of top secret messages. Ramp space for College Eye airplanes was at a premium, ground crew had to borrow tugs to tow the EC-121Ds, and other units helped themselves to any task force ground equipment left unattended.

In these circumstances, reported a College Eye senior noncommissioned officer, "all maintenance personnel became experts at locating and returning misplaced equipment." They also enlisted the aid of petty officers in a Navy detachment that flew Constellation transports, and the sailors proved "a valuable source of tools and parts." Whatever maintenance took place at Tan Son Nhut resulted from "personal contact of the NCOs with their counterparts on the base or in the Navy detachment." Luckily, mechanics at the task force's main support base on Taiwan performed all except minor repairs. [43]

When Col. Robin Olds, commander of the 8th Tactical Fighter Wing at Ubon, learned that College Eye was to move to his base, he made it clear to the task force commander, Lt. Col. Waldo W. Peck, that he did not like the idea of having EC-121Ds muscle their way in among his Phantoms, taking up ramp space and complicating supply problems. Seventh Air Force was determined, however, to reduce congestion at Tan Son Nhut, and the higher headquarters as always, prevailed. College Eye moved to Ubon in February 1967. Colonel Olds accepted the decision and proved a gracious host. More important, the colonel's pilots became interested in what the EC-12lDs could do for them and took part in planning sessions that led to the shift from MiG warning to fighter control. [44]

Not long after the task force began flying from Ubon, an incident occurred that resulted in what one of its officers called "a complete overhaul of the task force spirit. " An inspection party headed by Lt. Gen. James W. Wilson, Commanding General, Thirteenth Air Force, passed the parking area where the College Eye planes stood. Lieutenant General Wilson, with lesser officers in his wake, boarded one of the EC-121Ds. According to SMSgt Harold Rauback, who stood ramrod straight by the entrance, the general declared that the task force had the "sorriest, raunchiest, and most disgraceful machines he had ever seen. " He told the sergeant he would return within three weeks, at which time he expected to find the planes spotless.

EC-121s at Udorn. These are, however, EC-121Rs used to monitor the Igloo White sensors dropped on the Ho Chi Minh trail EC-121Rs are often misidentified as Ds or Ts despite the Rs lacking the dorsal height finding radar hump. Only EC-121Rs were camouflaged; EC-121Ds and EC-121Ts were not. Source: U.S. Air Force.

An orgy of cleaning ensued, during which everyone from Lieutenant Colonel Peck to the newest administrative clerk wielded brushes, soapy water, and rags. Anger gave way to humor, and humor became pride, as a sense of shared labor inspired the men of the task force. "I seriously doubt, " said one participant in the cleanup, "we would have scored the successes we did in the following year, " if it had not been for Lieutenant General Wilson's visit.

Colonel Olds gave the task force everything it needed to clean up the planes, and thanks to his cooperation the Thirteenth Air Force commander was able to inspect the EC-121Ds and pronounce them passably neat. Despite the colonel's initial reluctance, Ubon proved an infinitely more satisfactory base than Tan Son Nhut. College Eye received adequate ramp space and maintenance facilities, air conditioned offices, and vehicles for its exclusive use. If those in charge at Ubon were accommodating, so too in a different way was the local populace, for the task force venereal disease rate increased dramatically. [45]

Ubon was primarily a fighter base, however, and College Eye soon found itself part of a complicated series of moves that brought the task force temporarily to Udorn, also in Thailand, so that an F-4 unit could take up residence at Ubon. Meanwhile, construction was progressing at Korat air base, Thailand, which would become the EC-121D operating base. The shift to Udorn took place in July 1967, and the move to Korat in October. [46]

The crowded temporary base at Udorn lacked the maintenance facilities available at Ubon but nevertheless seemed a "very pleasant base overall. " When the task force arrived at Korat, not all the building was finished, but the task force commander, Colonel Davidson, was able to report that reaction to the new base was "enthusiastic." He added that the relationship between his unit and the host 388th Tactical Fighter Wing "exceeds all expectations." [47]

Acceptance and Departure

Late in 1967, everything seemed to fall into place for the College Eye Task Force. Morale had soared since the plane-washing affair in July. Korat was proving the best in a series of bases. Finally, F-4 crews were accepting the airborne controllers as full partners in the air war over North Vietnam.

This acceptance came at a critical time, for enemy interceptors were becoming increasingly aggressive. By the end of January 1968, General Momyer was expressing concern "over the advantage North Vietnamese MiGs enjoy over our aircraft by virtue of their GCI [Ground Control Intercept]." College Eye responded to the MiG challenge by assisting in the destruction of two enemy interceptors and the possible downing of a third, all within one week. [48]

The enemy, too, began to realize the growing effectiveness of College Eye. Late in January, General Momyer warned that North Vietnamese fighters might attack the EC-121Ds and other support planes. Scarcely had he uttered this warning than MiGs drew menacingly near Rivet Top as it was returning to Thailand after a patrol over the Gulf of Tonkin. Throughout February, hostile fighters approached EC-121Ds on Charlie station, and on one occasion a MiG came hurtling across the Laotian border, forcing the Connie to turn away as the fighter closed to within 25 nautical miles. Seventh Air Force met aggressiveness with caution, ordering the

EC-121Ds to remain south of the 19th parallel except during those hours when strikes were specifically scheduled to take place. [49]

Even as it helped meet this latest MiG challenge, College Eye reverted briefly to its air defense mission in response to reports of Russian-built biplanes operating near the demilitarized zone. For approximately 2 weeks in February 1968, the task force manned an additional orbit over the Tonkin Gulf off the North Vietnamese town of Dong Hoi. The EC-121s used search radar to cover an area from south of Da Nang northward beyond Vinh but did not detect the reported intruders. [50]

The capture of an American TACAN transmitter situated atop a mountain near the southern end of the Plain of Jars posed a potential threat to College Eye operations over Laos, because the EC-121Ds operating there relied on a navigational signal broadcast from this station to confirm their orbit. But actually, the loss of this checkpoint had little impact on task force operations, for President Johnson's ban on bombing north of the 20th parallel went into effect shortly after the site was captured, and College Eye pulled its aircraft southward to cover North Vietnam's panhandle. The new Laos orbit was east of Vientiane on the Laos-Thailand border, about 120 nautical miles south of the old position. Similarly, the overwater station had moved southward a distance of 60 nautical miles to a point seaward of Vinh. [51]

After 1 April 1968, College Eye support fighter-bomber missions over southern North Vietnam and parts of Laos, doing much the same work it had earlier. Airborne controllers were now responsible for enforcing a new buffer zone that reflected the bombing restrictions. In addition, a Rivet Gym aircraft kept the demilitarized zone under radar surveillance during June, searching unsuccessfully for re- ported enemy helicopter activity. [52]

The bombing halt, effective 1 November brought Rolling Thunder to an end, permitting College Eye to abandon the Tonkin Gulf orbit and operate exclusively over Laos. As far as North Vietnam was concern, emphasis shifted from the destruction of enemy aircraft to the tracking of friendly ones to prevent border violations. According to Col. Floyd M. McAllister, who commanded College Eye during the year ending in June 1970, the "most critical equipment for mission performance were the APS-95 search radar in the IFF [Identification Friend or Foe] mode, the APX-49 interrogator and the GPA-122 Decoder." The QRC-248, essential during Rolling Thunder, was of little value in the absence of radar-controlled North Vietnamese interceptors. Said Colonel McAllister: "Our detection of enemy aircraft by means of the QRC-248 was reduced as time went on, and toward the end of our tenure in SEA the QRC-248 was almost useless." [53]

The initial College Eye tour of duty ended on 29 June 1970, when the last of its planes left Korat for Itazuke air base, Japan. The organization departed at a time when the United States was reducing the number of Air Force support units in Southeast Asia. The legend, unsupported by evidence, has arisen that the withdrawal was an error by some budget analyst, trying to economize by transferring Detachment 1, 552d Airborne Early Warning and Control Wing, not realizing that this was the College Eye task force he was banishing. [54]

By the time College Eye left Southeast Asia, the tour of duty for task force members was 140 days. The officers and men who flew the Pacific with the original Big Eye detachment had returned to the United States after 90 days, but the Aerospace Defense Command soon found that it could not provide replacements so frequently. The long-standing Southeast Asia commitment disrupted the command's personnel policies, caused some dissatisfaction with the frequency of overseas tours, and led to temporary shortages of duty controllers and aircraft maintenance men. [55]

Although the withdrawal from Southeast Asia apparently was planned rather than accidental, the task force could not be spared for long. Four of its planes returned to Korat in September 1970 and spent 76 days supporting air operations over Cambodia. Again, in March 1971, College Eye dispatched planes to assist in the Cambodian enterprise. These left Thailand in July, but another contingent, flying the improved EC-121T, arrived in December, so that College Eye was on hand when North Vietnam launched its spring offensive against the South. [56]

Fitted out to function as the airborne component of an automated command and control network, the EC-121T was no stranger to the Vietnam war when it replaced the College Eye C-121Ds. In the summer of 1970 and in the spring of 1971, the T model had undergone combat evaluation in South-east Asia. Also, two of the planes, manned by volunteers under command of Lt. Col. John B. Mulherron, had taken part in the November 1970 attempt to rescue prisoners of war believed held at Son Tay, North Vietnam. [57]

During the Son Tay raid, the role of the EC-121Ts and its airborne replacement, was to control the F-4 fighter cover at altitudes below 7000 feet, where the curvature of the earth prevented coverage by shipboard equipment in the Gulf of Tonkin. En route to the station from which it was to perform the fighter control mission, one of the EC-121Ts developed an oil leak and had to turn back. The other plane was orbiting above the gulf when the crew discovered the recognition gear would not work. Since it could interrogate neither friendly nor enemy identification transponders, fighter control was impossible. The EC-121T did maintain radar surveillance, however, in order to provide MiG warning, but enemy interceptors did not intervene. [58]

Airborne Controllers in the 1972 Fighting

The North Vietnamese onslaught against the South early in 1972 increased the burden on College Eye air crews. Before the enemy's spring offensive, four EC-121Ts flying out of Karat had manned a Laos orbit from which they maintained surveillance over air operations in the northern part of that kingdom. On 1 March 1972, while carrying out this task, a Connie piloted by Capt. Walter Collins, with Capt. Joseph G. Euretig as senior director, helped an F-4D, manned by Lt. Col. Joseph W. Kittinger, Jr., and lst Lt. Leigh A. Bogden, down a MiG-21, the first successful interception controlled by College Eye since its return to South-east Asia in December 1971. During Freedom Train and Linebacker, kills of this sort became commonplace, as the task force increased to seven planes, reestablished a Tonkin Gulf station, and assumed a role similar to that it was performing when Rolling Thunder came to an end. [59]

During Linebacker the assigned EC-121Ts arrived at the Laos and Gulf of Tonkin stations before the day's attacks began. As the fighter- bombers and their escorts approached, the airborne controllers identified each flight, established radio contact, then followed transponder returns throughout the mission. From the crowded operations center inside the Connie, men of College Eye observed the replies of North Vietnamese transponders, sometimes triggering the devices, issued MiG warnings, and shifted patrolling Phantoms to meet the attacks. In addition, the task force issued collision warnings and helped pilots find tankers during refueling emergencies. [60]

Like the EC-121Ts, some Navy and Air Force F-4s could now intercept and decode MiG transponder returns, but they too still profited from College Eye surveillance. Crews of the specially equipped Phantoms, although able to identify a MiG at ranges beyond 100 nautical miles, had to have visual confirmation, unless some control agency such as College Eye could confirm the identification electronically. If the F-4 received this verification, it could launch an air-to-air missile without making visual contact. [61]

July 1972 proved an eventful month for the task force. On the 8th, Capt. Michael Edwards guided Captains Steve Ritchie and Charles B. Debellevue, pilot and weapons system officer of an F-4E, into position to shoot down a pair of MiG-21s But, unfortunately, success did not always crown College Eye's endeavors during that month. Late in July, the possibility was raised at a Seventh Air Force tactical conference that an error by an EC-121T controller had resulted in one F-4 lost and another damaged. His scope inundated by friendly and hostile returns, a tired controller may have used the wrong call sign so that a combat air patrol approaching Hanoi from the west reacted to a MiG alert intended for a chaff flight departing east of the city. [62]

The establishment of a weapons control facility at Nakhon Phanom, Thailand, caused a readjustment of College Eye responsibilities. The new installation. called Teaball went into action on 30 July, launching "an all out offensive against the MiGs." Presumably using equipment similar to that installed in the EC-121Ts, Teaball controllers picked up the enemy moments after he took off and tried to direct the F-4 fighter cover into position to prevent attacks on Linebacker strike forces. [63]

When Teaball began operating, its principal job was to provide information to controllers in College Eye EC-121Ts and on board the Navy radar ship, nicknamed Red Crown, in the Tonkin Gulf. Airborne controllers issued instructions to the combat air patrol, while Red Crown handled the strike force. In the event of radio or radar failure, the EC-121T yielded its responsibility to the shipboard controllers. Teaball, Under this arrangement, commenced tracking the interceptors as soon as they took off and radioed the data, via an automatic airborne relay, to the College Eye controllers, who repeated it for the Phantom crews.

Placing an EC-121T, which lacked automatic radio relay equipment, athwart the line of communication between Teaball and the patrolling Phantoms proved a mistake and within 3 weeks new procedures were put into effect. Beginning 20 August, College Eye yielded primary control of fighter patrol as soon as MiGs became airborne. At that moment Teaball took over, communicating directly with the fighter cover and also issuing warnings to chaff or strike forces when MiG attack was imminent. The EC-121T would take over should the radio link between Nakhon Phanom and the combat air patrol fail. [64]

Because the EC-121T carried equipment that picked up enemy identification signals, the controllers on board could acquire life-or-death information. As a result, they continued to watch the unfolding air battles, even though Teaball was in charge, interrupting to warn the combat air patrol of any rapidly developing threat that ground controllers might have missed. By the end of August, however, a new directive had restricted College Eye's initiative, specifying that: "Disco [the EC-121T call sign] may interject to pass information on IFF/ EIFF [Identification Friend or Foe /Enemy Identification Friend or Foe] only if this information would aid Teaball control. The primacy of Teaball controllers when MiGs were aloft was thus confirmed. [65]

The next change in College Eye duties occurred on 14 September, when it swapped primary responsibilities with Red Crown. The airborne controllers took over the chaff flight, the strike force, the follow-up photo reconnaissance flight, and their escorts (such as Iron Hand); Red Crown assumed responsibility for handling the combat air patrol when MiGs were not airborne. [66] Teaball took charge whenever enemy interceptors appeared but yielded to Red Crown when the Navy controllers decided "their air picture permits close control of the CAP [Combat Air Patrol] force. " After surrendering control to Red Crown, Teaball could "interject essential information when it will aid in the intercept." [67] If Teaball went out of action, Red Crown took its place, with Cisco backing up the Navy facility. Once again, the change seemed to be an attempt to entrust control of the combat air patrol to a facility with dependable communications. [68]

On 6 October, under the new procedures, Disco issued a MiG warning to an Iron Hand flight escorting a Linebacker strike force. The two F-4Es in the flight turned to dispose of the threat, but one of them had to dive into a valley in order to evade a MiG-19. The enemy pilot pursued but was unable to pull up and crashed to his death. After assisting in this unorthodox kill, College Eye reverted to more normal tactics on the 8th, when lst Lt. Michael Clifton alerted Maj. Gary L. Retterbush and Capt. Robert H. Jasperson, members of an F-4E crew, who shot down a MiG-21. [69]

For the balance of the air war against the North, the College Eye Task Force employed radar and recognition gear in tracking night-flying F-111As, controlled fighter cover for B-52s attacking the passes leading from North Vietnam into Laos, and collaborated with Teaball and Red Crown during Linebacker II. The orbit assigned to the EC-121Ts during the last phase of the campaign apparently limited the usefulness of the airborne controllers. Task force officers complained that their overland station was too low to obtain satisfactory coverage of the B-52 raids around Hanoi and Haiphong. [70]

Summing Up

The arrival of the Big Eye task force in Southeast Asia stirred little enthusiasm among the units already there, and 2d Air Division was at first reluctant to use the EC-121Ds on their logical mission, since they seemed suicidally vulnerable to enemy fighters. This fear proved groundless, however, for MiGs failed to down any of these planes or the T models that supplanted them. Even though the specially equipped aircraft soon demonstrated their value in issuing MiG warnings, some pilots continued to resent the border warning missions performed by Big Eye and College Eye, an activity that could result in a reprimand for an incautious airman. During 1967, however, the task force received equipment that enabled it to track the responses from Russian-built identification transponders. No longer content with merely monitoring air traffic and issuing appropriate warnings, task force controllers began directing F-4s against enemy interceptors. This new equipment contributed to most of the 23 confirmed and 2 probable MiG kills in which the task force participated from 1965 through 1972. [71]

Despite this success, neither the EC-121D nor the more advanced T model fully satisfied the needs of fighter pilots. For example, Col. George W. Rutter, who commanded the 366th Tactical Fighter Wing during Linebacker, complained that Red Crown, Teaball, and Disco had failed to give "accurate, real-time information on MiG activity and precise, positive control to the CAP [Combat Air Patrol) and escort flights so as to cope with the MiG threat." The EC-121T, he believed, had "radar system limitations that preclude effective radar coverage of the target area where the MiG defenders have the advantage of excellent GCI [Ground Control Intercept) control. "

Although Teaball had proved a "valuable aid, " he maintained that "the only long-term solution to this problem" lay in development of a replacement for the EC-121T "that can provide look-down radar coverage of the target area combined with positive control over counterair fighters operating in that area." [72]

The radios installed in the EC-121 T represented an improvement over those in the first D models but nevertheless imposed limits on College Eye effectiveness. From the 1972 Laos orbit, the ultra high frequency signal tended to dissipate in the vicinity of Hanoi and seldom reached Red Crown without help from relay equipment. The Connie, moreover, did not have an automatic relay on board, an obvious disadvantage in passing information between Teaball and the Navy control center. [73]

The significant contributions made by officers and men of Big Eye and College Eye were the more striking because of equipment they had to use. As one of their number, Maj. Lowell J. K. Davis, has written. "the dedication and determination of the EC-121 radar crews" offset the "inherent limitations" of what he termed "a relatively obsolete airframe and a weapons system designed for a less demanding role." Thanks to continual modification and skilful operation, the aging equipment formed an essential link in a control network that enabled American fighters to cope with a capable and determined foe.

74

GLOSSARY

A-1 Douglas Skyraider attack bomber, powered by a single radial engine, built in single-seat and two place versions, and flown by the Navy and Air Force.

A-4 McDonnell Douglas Skyhawk, a single-seat attack bomber powered by one turbojet engine, and used by the Navy and Marine Corps.

A-6 Two-place, twin turbojet, all-weather attack plane built by Grumman for the Navy and Marine Corps.

A-7 Single-place attack plane, powered by a single jet engine, built by Ling- Temco- Vought for the Navy and Air Force.

AAA Antiaircraft artillery

Acquisition Radar

Detects targets at a range of about 100 nautical miles and tracks them to within range of fire control radars.

AD Air Division

ADC Aerospace Defense Command (Air Defense Command prior to 1968)

AEW&C Airborne Early Warning and Control

AF Air Force

AFB Air Force Base

AFM Air Force Manual

AFP Air Force Pamphlet

AFSC Air Force Systems Command

AFSSO Air Force Security Service Office

AGM Air-to-ground missile

AGM-45 Shrike air-to-ground missile; homes on radar transmitters.

Glossary

AGM-78	Standard ARM air-to-ground missile; homes on radar transmitters and incorporates improvements over the older Shrike.
AIG	Address Indicator Group
ALQ-51	Self-protection jammer; deceives enemy radar operators by broadcasting a false radar return.
ALQ-71	Redesignation of the QRC-160A-l, See QRC-160.
ALQ-87	Self-protection pod, originally designated QRC-160- 8. Entered service late in 1967
ALQ-94	Deception jammer carried by the General Dynamics F-111.
ALQ-101	Self-protection pod, which saw service in 1972; per- forms deception and noise jamming.
ALQ-119	Self-protection pod designed to replace the ALQ-71 and ALQ-87; capable of noise and deception jamming.
ALR-18	Jamming transmitter designed for use against air- borne radar but employed by Linebacker II B-52 1 s against the T-8209 radar.
ALT-6	B-52 jamming transmitter; being replaced during 1972 by the ALT-22.
ALT-22	Jamming transmitter in- stalled in all Linebacker II B-52Ds and in some B-52Gs.
ALT-28	Jamming transmitter employed by Linebacker II Stratofortresses against track-while-scan radar and the SAM guidance beacon.
APGC	Air Proving Ground Center
APP	Appendix
Apr	April
APS-95	Search radar installed in the EC-121D
APX-49	Recognition set used to identify friendly aircraft.
ARC-37	Ultra high frequency radio.
ARC-109	Radio that replaced the RC-27 in the EC-121D during 1968.
Arc Light	B-52 operations in Southeast Asia.
ARM	See AGM-78.
Atch	Attachment
Aug	August

AW	Automatic weapons.
AWC	Air War College.
B-17	Flying Fortress; Boeing- built four-engine bomber that saw extensive service in World War II.
B-29	Superfortress; four-engine bomber built by Boeing, which carried the war to Japanese cities in 1945, and served in the Korean conflict.
B-50	Boeing Superfortress; replaced the B- 29 and saw combat in Korea.
B-52	Stratofortress; swept-wing strategic bomber built by Boeing and powered by eight turbojet or turbofan engines.
B-57	Martin-built version of the British Canberra twin-jet medium bomber.
B-66	See EB-66
B-427Z	Teamwork. See T- 8209
Barlock radar	Ground control intercept radar.
Barrage jamming	The form of electronic jamming in which the operator distributes power over a wide frequency band.
Barrel roll	Aerobatic maneuver in which the plane is rolled while simultaneously revolving around a fixed axis, so that it follows a cork-screw flight path.
BASS	Acronym for Bistatic Aided Strike System, an attempt to use radar detection gear on board a Rivet Top EC-121 to direct Wild Weasel aircraft against SAM sites.
Beacon	See guidance beacon.
Big Eye	See College Eye.
Big Look	EC-121M; Navy version of the EC-121.
Bk	Book
Brig Gen	Brigadier General
Burn through	Juncture at which radiated power of the radar transmitter overcomes the jamming signal generally at a range of 8 to 10 nautical miles, enabling the operator to isolate the target.
C-130	Lockheed Hercules; four-engine turboprop transport.

Glossary

CAP	Combat Air Patrol.
Capt	Captain
CETF	College Eye Task Force.
CG	Commanding General.
Ch	Chapter.
Chaff	Radar reflectors dispensed from aircraft to confuse the picture on enemy scopes.
Chaff Bomb	Leaflet dispenser, containing an explosive charge, modified to scatter chaff.
CHECO	Contemporary Historical Evaluation of Combat Operations.
CIA	Central Intelligence Agency
Cigar	World War II countermeasures equipment that jammed radio communication among German interceptor pilots and ground controllers.
CINCNORAD	Commander in Chief, North American Air Defense Command
CINCPAC	Commander in Chief, Pacific.
CINCPACAF	Commander in Chief, Pacific Air Forces.
CINCSAC	Commander in Chief, Strategic Air Command.
Clutter	See noise.
co	Commanding Officer.
Co	Company
Col	Colonel
College Eye	Task force of EC-121s; provided early warning of air attacks from the North. Officially, Detachment 1, 552d Airborne Early Warning and Control Wing.
Comfy Boy	Reports on electronic countermeasures activity in Southeast Asia.
Comfy Coat	Reports on electronic counter-measures activity in Southeast Asia.
Connie	Lockheed EC-121 Constellation.
Constellation	See EC-121.
Corona Harvest	Air Force project to collect historical data on the South-east Asia war.

CSAF	Chief of Staff. U.S. Air Force.
DCS	Deputy Chief of Staff
Dec	December
Deception	Form of jamming in which the aircraft broadcasts a false radar return to mislead the defenders.
Demilitarized Zone	Buffer area established between North and South Vietnam.
Dep	Deputy
Dept	Department
Destroyer	See EB-66
Det	Detachment
Dg	Downgraded.
DIA	Defense Intelligence Agency.
Dir	Director
Disco	A call sign used by Air Force EC-121s.
Div	Division
Doc	Document
Down link	See guidance beacon.
Drone	A remotely controlled pilot- less aircraft.
EA-1	Electronic warfare version of the Douglas A-1 Skyraider.
EA-3	Electronic warfare version of Navy's Douglas Skywarrior twin-jet light bomber.
EA-6	Grumman A- 6 modified for electronic warfare.
EB-47E	Electronic countermeasure version of the Boeing Stratojet medium bomber, a swept- wing aircraft powered by six turbojet engines.
EB-66B	Electronic warfare version of the Air Force twin-jet RB-66 reconnaissance plane, descended from the same Douglas prototype as the Navy EA-3.
EC-121	Lockheed Constellation four-engine military transport converted for electronic warfare.
ECM	Electronic countermeasures.

Glossary

Ed	Editor
EF-10B	Two-place, twin-jet Douglas Skyknight equipped for electronic countermeasures missions and used by the Navy and Marine Corps.
EIFF	Enemy identification friend or foe.
et al.	And others
EWO	Electronic Warfare Officer
F-4	McDonnell Douglas Phantom, a two-place, twin-jet fighter bomber used by the Air Force, Navy, and Marine Corps.
F-100	Single-jet, single-seat North American Super Saber fighter-bomber, also employed by the Air Force in a two-seat version.
F-104	Lockheed's Starfighter, a single-place, lightweight Air Force fighter, powered by one turbojet engine.
F-105	Republic Thunderchief fighter-bomber, a single-jet aircraft built for the Air Force in single-seat and two-place models.
F-111	Air Force twin-jet, two-place, variable-sweep aircraft built by General Dynamics for all-weather operation.
Fan Song	Track-while-scan radar used to determine the azimuth and range to SAM targets.
Feb	February
Ferret	Electronic reconnaissance aircraft or mission.
Fingertip formation	One in which a flight of tour aircraft maintains an alignment resembling the fingertips of the human hand extended palm downward.
Fire Can	Soviet-designed radar used by the North Vietnamese to control antiaircraft guns ranging in size from 37-mm to 100-mm.
Flak	Antiaircraft fire or anti-aircraft guns.
Flying Boom	A method of refueling in which an operator on board the tanker maneuvers a fuel-carrying boom into a receptacle on the fuselage of the other aircraft.
Flying Telephone Pole	Guideline surface-to-air missile.
Freedom Dawn	B-52 attack against Bai Thuong airfield, near Thanh Hoa, North Vietnam, on 12 April 1972.

Freedom Porch B- 52 strikes against Haiphong North Vietnam, 15 April 1972.

Freedom Porch Bravo

Tactical strikes against the Hanoi-Haiphong area delivered on 6 April 1972.

Freedom Train Operation, lasting from 6 April through 8 May 1972, in which U.S. aircraft attacked targets between the demilitarized zone and the 20th parallel.

Freedom Train Bravo

B-52 attack against Vinh, North Vietnam, on 9 April 1972.

Freighter Captain B-52 strike against Thanh Hoa, North Vietnam, 21 April 1972.

Frequent Winner B-52 attack on Thanh Hoa, North Vietnam, 23 April 1972.

GCI Ground Controlled Intercept

Gen General

Gimp An evasive maneuver used by drones to frustrate fighter attacks.

GPA-122 Recognition equipment capable of automatically decoding the response from identification transponders carried by U.S. aircraft.

Guidance beacon Signal or signaling device that tells weapons controllers the trajectory of a surface-to- air missile.

Guided bombs Ordnance directed to the target by a member of an aircraft crew using either television or a laser beam.

Guideline Missile used with North Vietnam's SA- 2 system, it consists of booster and sustainer stages, measures 10.6 meters (35 feet) in length, and carried a 189.6 Kilogram (420-pound) high explosive warhead detonated by a radar proximity fuse.

Hat Rack Device that could recognize both MiG and SAM threats and trigger appropriate drone response.

Heat-seeking missile

Missile fitted with an infra-red device enabling it to home on the heat generated by jet engines.

HC-130P Tanker version of the Lockheed C-130

Hist History

Hq Headquarters

Glossary

Hunter- killer	Tactics which employ hunter aircraft, usually carrying special detection equipment, to locate targets for munitions-carrying killers.
Ibid	The same source cited immediately above.
IFF	Identification Friend or Foe
Intvw	Interview
Iron Hand	Originally hunter-killer operations against SAM sites, the term came to refer to radar suppression involving specially equipped aircraft.
Jamming Package	Standardized instructions prescribing jamming procedures for EB-66 electronic warfare officers.
Jan	January
JCS	Joint Chiefs of Staff
Jr	Junior
Jul	July
Jun	June
KC-135	Boeing Stratotanker, powered by four jet engines, which served in Vietnam as a radio relay and communications intelligence aircraft as well as a tanker.
Linebacker	Air Attacks, beginning 9 May 1972, against targets throughout North Vietnam. Carried out in conjunction with minelaying and a naval blockade, the operation was designed to isolate North Vietnam from foreign sources of military aid.
Linebacker II	Sustained B-52 attacks against targets in the Hanoi- Haiphong area, 18-30 December 1972.
LORAN	Long Range Navigation, a system of navigation in which position is determined by the time difference in the arrival of signals from ground stations.
Lt	Lieutenant.
Lt Col	Lieutenant Colonel
Lt Gen	Lieutenant General
Magnetron	Type of vacuum tube used to generate jamming power.
Maj	Major

Maj Gen	Major General
Mandrel	British World War II countermeasures transmitters that jammed German early warning radars.
Maneuver pod formation	More flexible version of the pod formation. It was used during 1972 to provide better visual coverage to the rear of the flight.
Mar	March
Mgt	Management
MiG	Symbol indicating that an aircraft is the product of A. I. Mikoyan and M. I. Gurevich, a team of Russian designers. Often used as a generic term for Russian-built fighters and erroneously written as MIG.
MiG-17	Swept-wing, single-place interceptor powered by a single turbo-jet engine.
MiG-19	First Soviet fighter-interceptor capable of supersonic speed, it was a single-turbojet, swept-wing aircraft with a one-man crew.
MiG-21	Short-range, delta-wing interceptor powered by a single turbo-jet engine and carrying a crew of one.
Mm	Millimeter
Msg.	Message
Nd	No date
No	Number
Noise Jamming	Signal designed to obscure the image on the radar scope; the interference or clutter that appears on the scope.
Normal jamming	Disruption of the Fan Song track-while- scan radar beam.
Nov	November
Np	No pagination.
NSA	National Security Agency.
NVN	North Vietnam
Oct	October
Ofc	Office
Opl	Operational
Ops	Operations

Glossary

p	Page
PACAF	Pacific Air Forces
PACOM	Pacific Command
Panoramic scan receiver	Electronic device used to determine the bearing to enemy radar.
passim	Throughout
Passive tracking	Radar operator tracks the source of jamming; he does not employ his tracking beam, since its return would be masked by the jamming signal.
Phantom	See F-4
Pod formation	Formation flown in such a manner that the jamming signals from the self-protection pods on the individual aircraft reinforce each other.
Post target turn	Change in heading that occurs after bomb release as the attacking aircraft begin their exit from the target area.
PP	Pages.
Probe and drogue	Aerial refueling method in which the fuel intake pipe -- the probe -- is inserted into a funnel-shaped receiver at the end of a hose trailing from the tanker.
Proud Deep Alpha	Series of strikes against airfields, petroleum storage facilities, and military barracks in North Vietnam, 26-30 December 1971.
Prov	Provisional
Proximity fuse	See radar proximity fuse.
Pt	Part
Pulse Repetition Frequency	Number of radar pulses generated per second.
QRC-128	Equipment, installed in the rear cockpit of some fighter-bombers or attack planes to jam radio communication between ground controllers and interceptor pilots.
QRC-160	Family of self-protection pods whose development began prior to the Vietnam war. The QRC-160A failed its combat test, but the QRC-160A-1 proved satisfactory.

QRC-248 — Device for interrogating the identification transponders carried by North Vietnamese fighters.

QRC-335 — Jamming device fitted to the Wild Weasel F-105G and capable of deception as well as noise jamming.

R&D — Research and Development

Radar Homing and Warning — Electronic equipment that warns of hostile radar tracking and indicated the general location of the enemy set.

Radar proximity fuse. — Fuse that detonates upon signal from a miniature radar transmitter in the device itself.

Ram air turbine — Turbine for which the motive power is a flight-generated stream of air.

RB-66 — See EB-66

RC-121 — See EC-121

Red Crown — Navy control ship positioned in the Tonkin Gulf and capable of controlling aircraft over North Vietnam.

RF-4 — Reconnaissance version of the McDonnell Douglas F-4.

RF-101 — Two-place twin turbojet reconnaissance plane built by McDonnell Aircraft.

Rivet Bounder — Deception jammer activated by the Fan Song guidance signal.

Rivet Gym — EC-121 modified to include the fighter control features of Rivet Top.

Rivet Top — Lockheed EC-121K fitted out as an airborne command post for the air war over North Vietnam.

Rolling Thunder — Air war conducted against North Vietnam from March 1965 through October 1968.

Route Package 2 — Northern half of the panhandle region, one of seven carefully defined portions of North Vietnam that came under attack during Rolling Thunder.

Route Package 3 — Segment of North Vietnam lying just north of Route Package 2.

Rprt — Report

RSI — Research Studies Institute.

Glossary

Ryan's Raiders	F-105 unit that conducted night harassment against the North; so designated because of Gen. John D. Ryan's interest in the project.
S	Secret
SA-2	Soviet-designed surface-to-air missile system incorporating the Fan Song radar, the Guideline missile, a fire control computer, and a target acquisition radar, plus launchers and motor transport.
SAB	Scientific Advisory Board.
SAC	Strategic Air Command
SAM	Surface-to-air missile
Scrambler	Coding system that distorts radio or telephone conversation as it leaves the transmitter and returns it to an understandable form at the receiver.
SEA	Southeast Asia.
Self-protection pod	Aerodynamically shaped jamming device carried externally by tactical aircraft.
Sep	September
Short Stirling	Product of Short Brothers Aircraft, this four engine British heavy bomber saw extensive service in World War II.
Shrike	See AGM-45.
Skyhawk	See A-4
Skyknight	See EF-10B
SOD-57	Identification equipment, originally intended to aid air traffic controllers, used to direct North Vietnamese MiGs.
Sortie	Takeoff and landing by a single aircraft.
Sparrow	Solid-propellant, radar-guided air-to-air missile developed by the Raytheon Company.
Spec Comm	Special Communications
Special jamming	Disruption of the signal from the guidance beacon, or down link, mounted on the Guideline missile.
Spoon Rest	North Vietnamese acquisition radar frequently used in conjunction with Fan Song.

Spot Jamming	Form of electronic jamming in which all available power is concentrated on a narrow frequency range.
Squawk	To use the identification gear carried by individual aircraft.
SR0-1	Identification equipment carried by early-model MIG interceptors.
Standard ARM.	See AGM-78
Stand-off jamming	Jamming from long range, as the countermeasures aircraft remains beyond reach of surface-to-air missiles.
Stratofortress	See B-52 (The B-52 is also known as Miss Buffy and The Gray Lady)
Subj	Subject
Superfortress	See B-29, B-50
Super Sabre	See F-100.
Sweep jamming	Form of electronic jamming in which a concentrated beam is swept over a wide frequency range.
Sweep modulator	Component of a self-protection pod that introduces random bursts of energy into the jamming barrage.
SW	Strategic Wing
T-8209	Radar signal employed late in the war to direct North Vietnamese surface-to-air missiles and antiaircraft guns. Also known as B-427Z --Teamwork.
TAC	Tactical Air Command
TAWC	Tactical Air Control Center, North Sector Control agency established in 1966 for the air war over North Vietnam.
Teaball	Tactical Air Warfare Center. Weapons Control Center at Nakhon Phanom, Thailand.
Terrain masking	Protection afforded by ridges or similar geographic features against radar detection.
TEWS	Tactical Electronic Warfare Squadron.
TFS	Tactical Fighter Squadron
Thud	Thunderchief. See F-105D.
Thud Ridge	Ridge line used by Air Force F105 's for terrain masking as they approached the Hanoi area from the northwest.

Glossary

Thunderchief	See F-105
TF-77	Task Force 77, the U. S. carrier task force operating off the Vietnam coast.
Tonkin Gulf incident	Attack on 2 July 1964 by North Vietnamese patrol boats on the U.S. destroyer Maddox.
Track-while-scan	Feature of the Fan Song radar enabling it to detect additional aircraft while tracking a specific target.
Triple A	See AAA.
Trolling	Missions in which electronic warfare aircraft attempt to trigger enemy radar in order to locate or attack the transmitters.
TRS	Tactical Reconnaissance Squadron.
TRW	Tactical Reconnaissance Wing
TS	Top Secret
U	Unclassified
U-2	Lockheed's single-turbojet single-place, high-altitude reconnaissance plane.
USAF	U.S. Air Force.
USAFE	U. S. Air Forces, Europe
USAFSS	U.S. Air Force Security Service
USMC	U.S. Marine Corps
USN	U.S. Navy
VHAW	Vector homing and warning -- Electronic equipment that warns of enemy radar tracking and furnishes a bearing to the hostile transmitter.
Vol	Volume
Voodoo	See RF-101
Wg	Wing
Wild Weasel	F-100F, F-105F or G. or F-4C aircraft fitted with radar homing and warning gear for the neutralization or destruction of radar-controlled weapons.
Window	British World War II term for chaff.
z (Zulu)	Greenwich Mean Time

BIBLIOGRAPHICAL ESSAY

This study is based on material from six broad categories: messages; published works; Air Force efforts to document the war, including Projects CHECO and Corona Harvest, and monographs prepared by historical offices at Air Force headquarters and at the major commands; tactical manuals; reports; and recurring histories submitted by commands, numbered air forces, wings, task forces, and squadrons. A seventh category, special intelligence dealing with the effectiveness of countermeasures over North Vietnam, was left untouched because of access restrictions.

Because the author did not have access to this essential material, the Chief, Histories Division, Office of Air Force History, Mr. Carl Berger, arranged for persons having the necessary security clearance to review the draft manuscript and make sure that this account was not contradicted by information unavailable to the person who wrote it. Extremely helpful comments were attached to letters sent to Mr. Berger by these Air Force officers: Col. Floyd A. McLaurin, Commander Air Force Electronic Warfare Center; Col. Robert L. Rodee, Chief, Tactical Division, Directorate of Operations, Headquarters USAF; Col. Edgar A. Gill, Jr., Chief, Strategic Forces I C3 Division, Directorate of Programs, Headquarters, USAF; Col Edward L. Scott, Director of E W Operations, Head- quarters Tactical Air Command; Maj. Allan P. Botticelli, Analysis and Evaluation Group, Assistant Chief of Staff, Studies and Analysis, Headquarters USAF; Maj. John P. Shmoldas, Executive Officer, Directorate of Reconnaissance and Electronic Warfare, Headquarters, USAF; Col. Robert S. Johnson, Assistant Deputy Chief of Staff, Operations, Aerospace Defense Command; Maj. Jesse P. Wiggins, Directorate of Operations, Tactical Air Command; and Col. Peter Tsouprake, Electronic Systems Division, Air Force Systems Command.

The author did consult those U.S. Air Force Security Service reports to which he had access and also relied upon a top secret monograph, Electronic Warfare in SEA, by James E. Pierson, a Security Service historian. In dealing with electronic countermeasures, or any other facet of electronic warfare, the Air Force historian cannot avoid the security problem. If a study is to be readily available throughout the Air Force, it cannot probe the topic too deeply. The inevitable result has been a succession of histories that deal superficially with the subject, sometimes actually misleading the reader who is not already familiar with the more arcane aspects of electronic warfare.

Although helpful in themselves, Project CHECO and Corona Harvest reports were less valuable than the documents and submissions gathered for their preparation. Especially helpful were the "inputs" to Corona Harvest by Pacific Air Forces and Aerospace Defense Command. The recurring histories submitted by participating units provided data on operations and maintenance. Forwarded with them were thousands of exhibits and supporting documents, including key messages and many of the compilations of tactical doctrine used in this study. One history in particular was remarkably well done--the account of the 307th Strategic Wing (Provisional), October-December 1972.

The important data on Linebacker II was collected by John Greenwood in his chronology of SAC participation in that operation.

The appendix dealing with fighter control could not have been written without the excellent history prepared in 1969 by Capt. Richard M. Williams, USAF. The Hist of College Eye, April 1965 - June 1969, by Grover C. Jarrret, yielded a great deal of information as did the CHEC0 report on College Eye written by Capt. Carl W. Reddel, USAF.

For insights not found in documents, the author turned to the Air Force collection of oral history interviews and the end-of-tour reports submitted for Project Corona Harvest. Cited in this study are some 10 interviews and 15 end-of-tour reports. Of these, the most useful were the interviews with Col. Floyd McAllister, Lt. Col. Robert E. Belli, and Lt. Col. Merlyn Dethlefsen, and the reports submitted by Col. Robin Olds, Col Ian D. Rockwell and Marine Capt. John L. Pycior.

NOTES

Chapter I

1. Public Papers of the Presidents of the United States (Washington: Ofc of Federal Register, National Archives and Records Service, 1966), Lyndon B. Johnson, 1965, Bk I, p 208; Capt Melvin F. Porter, USAF, Air Tactics against NVN Air/Ground Defenses (TS) (Hq PACAF, Project CHECO, 27 Feb 67), pp 8-9, cited hereafter as Porter, Air Tactics.

2. Msg (S), 45th TFS to 2AD, subj: Aircraft Mishap Rprt, 241505Z Jul 65, supporting doc no 5 to Porter, Air Tactics.

3. Based on Study (S), Dep for Foreign Technology, APGC, AFSC, Opl Data on the SA-2 System, Mar 66, cited hereafter as Data on the SA-2 System, Mar 66; JCS Prong Tong Study (TS), Surface to Air Missile Problem in North Vietnam, 15 Oct 65, vol II, app 1 to annex III, pp 1-13, cited hereafter as JCS Prong Tong Study.

4. Francis Gary Powers with Curt Gentry, Operation Overflight: The U-2 Pilot Tells His Story for the First Time (New York: Holt, Rinehart, Winston, 1970) pt II passim; "To Them: Sorrow, Gratitude, Pride, " Air Force and Space Digest, Dec 62, p 21; Victor Marchetti and John D. Marks, The CIA and the Cult of Intelligence (New York: Alfred A. Knopf, 1974), p 310.

5. Msg (S), 7AF to TAC Dir/Ops, subj: Electronic Warfare Summary, 310425Z Dec 72.

6. Msg (S), 7AF to CSAF, subj: F-111 Altitudes, 171214Z Oct 72; Col A. A. Picinich et al. , The F-111 in Southeast Asia, September 1972-January 1973 (S), (Hq PACAF, Project CHECO, 21 Feb 74), p 30, cited hereafter as Picinich, The F-111.

7. Hist (S), 388th TFW, Jul-Sep 72, vol I, pp 37-38.

8. Atch 1 (S) to ltr, Col Floyd A. McLaurin, USAF, Commander, AF Electronic Warfare Center to Ofc of AF Hist, subj: Review of Historical Manuscript, 11 Mar 76, cited hereafter as McLaurin ltr.

9. Ibid.

10. JCS Prong Tong Study, vol II, app 2 to annex C, pp 1-6' Study (S), North Vietnamese Air Defense System, atch to Navy Dept Vietnam Appraisal Briefing, 19 May 67, cited hereafter as North Vietnamese Air Defense System.

11. North Vietnamese Air Defense System; Data on the SA-2 System, Mar 66.

12. JCS Prong Tong Study, vol IL app 2 to annex C, pp 1-6.

13. North Vietnamese Air Defense System.

14. JCS Prong Tong Study, vol IL app 2 to Annex 3, pp 1-6.

15. Project Corona Harvest Input, Support: Electronic Warfare, 1965-1968 (TS), Hq PACAF, p 23, cited hereafter as Corona Harvest, Electronic Warfare.

16. Porter, Air Tactics, pp 2-4; John W. R. Taylor, ed, Combat Aircraft of the World from 1909 to the Present (New York: G. P. Putnam's Sons, 1969), pp 587-589, cited hereafter as Taylor, Combat Aircraft.

17. End-of-Tour Rprt, 30 Nov 69-11 Dec 70 (S), Col Morris E. Shiver, USAF, CO 42d TEWS, p 6.

18. R. C. Guenther, "North Vietnamese SA-2 System, Development and Role," (S), Defense Intelligence Digest, Feb 71, p 7.

19. "History of Air Defense in North Vietnam" (S), PACOM Intelligence Digest, May 70, pp 35-42; Study (TS), Dir/Ops, Hq USAF, Rolling Thunder-Linebacker: A Preliminary Comparative Analysis, Jun 72, Tabs 9-12, cited hereafter as Rolling Thunder- Linebacker Comparison; Chronology (TS), SAC Participation in Linebacker II (SAC, 12 Aug 7' 1, p 26, cited hereafter as SAC Linebacker II Chronology.

20. History of Air Defense in North Vietnam" (S), cited above; Southeast Asia Military Fact Book (S), DIA-JCS, Jul 66, pp 111-112.

21. SAC Linebacker II Chronology, pp 26, 169, 193.

22. Sir Charles Webster and Noble Frankland, The Strategic Air Offensive against Germany (London: Her Majesty's Stationery Ofc, 1961), vol II, pp 141-145, 202; vol III.1pp 150-151; Anthony Verrier, The Bomber Offensive (New York: Macmillan Co., 1969), pp 151, 292.

23. Hist (S), 376th Bombardment Wg, Nov 51, ch II; Dec 51, ch II; AFM 51-3 (S), Electronic Warfare Principles, 1 Nov 62, p 5:9.

24. Robert F. Futrell, Air Force Operations in the Korean Conflict, Jul – Jul (U) USAF Historical Study 127 (RSI, Air University, 1956), pp 77 -79.

25. AFM 51-3, 1 Nov 62, pp 5:9-10

26. Maj Wayne J. Clay, USAF, Electronic Countermeasures: Should We Reorganize Our Thinking? (S), (Air University: Air Command and Staff College, 1968), pp 30-31, cited hereafter as Clay Thesis.

27. Hist (S), TAC, Jan-Jun 57, vol I, pp 317-319; Draft (U), TAC Presentation for Scientific Advisory Board, 11 Sep 57, cited hereafter as TAC Presentation to SAB.

28. Hist (S), Dir Ops, DCS/Plans and Ops, Jul-Dec 63, pp 6-7.

29. TAC Presentation for SAB.

30. Ibid.; Hist (S), 41st TRS, Jan-Jun 66, doc 7; Final Rprt (S), Swamp Fox 1, 30 Jun 60, supporting doc no 2 to hist, 363d TRW, Jul-Dec 60.

31. Hist (S), TAC, Jan-Jun 57, p 325; Hist (S), 41st TRS, Jan- Jun 66, doc 7.

32. Hist (S), 363d TRW, Jan-Jun 59, p 22; Jul-Dec 59, p V:39; TAC Presentation to SAB.

33. Hist (S), TAC, Jan-Jun 57, vol I, pp 317 - 319.

Chapter II

1. Hist (S), 9AF, Jul- Dec 65, vol I, pp 51-53; Hist (S), PACAF, Jan- Dec 67, pp 66-67; Hist (S), 6460th TRS, 30 Jun-18 Sep 66, np; Hist (S), 355th TFW, Jul-Sep 67, vol I, p 15; USAF Mgt Summary: Operational Status (S), 20 Aug 65; Southeast Asia (S), 24 May 68; Msg (S), TAC to CSAF et al. subj: RB-66C Electronic Warfare Requirements, filed 27 May 65, exhibit 181 to Hist, 9AF, Jan -Jun 65.

2. Lt Col Robert M. Burch, USAF, Tactical Electronic Warfare Operations in SEA, 1962-1968 (S), (Hq PACAF, Project CHECO, 10 Feb 69), p 19, cited hereafter as Burch, Tactical Electronic Warfare.

3. Taylor, Combat Aircraft pp 492-496; Summary of Air Ops, Southeast Asia (S), PACAF Tactical Evaluation Center, 1 - 8 Jul 65, 9 - 22 Jul 65, 23 Jul - 5 Aug 65, cited hereafter as Summary of Air Ops, Southeast Asia.

4. Rprt of Overseas Tour (S), Capt John L. Pycior, USMC EF-10B crewman, 11 Mar 66, pp 2-17.

5. Burch, Tactical Electronic Warfare, p 19; Hist (S), 41st TRS, Jan-Jun 66, doc 7.

6. Corona Harvest, Electronic Warfare, pp 38-39.

7. Ibid., pp 39-40; Hist (S), 41st TRS, Jan-Jun 66, p 1:8.

8. Hist (S), 41st TRS, Jan-Jun 66, doc 7.

9. 7AFP 55-2 (S), Rolling Thunder Ops Handbook, 15 Jul 68, p 29, cited hereafter as Rolling Thunder Handbook.

10. Burch, Tactical Electronic Warfare, pp 33-34.

11. James E. Pierson, Electronic Warfare in SEA, 1965- 1968 (TS), (USAFSS, 1973), p 68, cited hereafter as Pierson, Electronic Warfare; Summary of Air Ops, Southeast Asia, PACAF Tactical Evaluation Center, 18 Feb - 3 Mar 66, pp 7:L-1 and 2.

12. Hist (S), 41st TRS, 18 Sep-31 Dec 66.

13. Corona Harvest, Electronic Warfare, pp 48-49; EB-66 Tactics Manual for SEA (S), 355th TFW, app I, pp 1-2; Hist (S), 355th TFW, Oct-Dec 67, vol I, pp 28-29.

14. Hist (S), 355th TFW, Oct-Lee 67, vol L pp 29-30; EB-66 Tactical Manual for SEA (S), 355th TFW, app I.

15. Hist (S), 355th TFW, Apr-Jun 68, vol I, p 56.

16. End-of-Tour Rprt (S), Col. Ian D. Rothwell, USAF, CO 41st TEWS, 10 Dec 67 – 4 Dec 68.

17. Ibid.

18. Hist (S), 41st TRS, Jan-Jun 67, doc 7; Hist (S), 355th TFW, Jul-Dec 67, vol I, pp 119-120; Rolling Thunder Handbook, p 29.

19. Atch 1 to McLaurin ltr; Corona Harvest, Electronic Warfare, pp 50-51.

20. Corona Harvest, Electronic Warfare, p 52; Atch (S) to ltr, Maj Allan P. Botticelli, USAF, Analysis and Evaluation Gp, Ofc of Asst Chief of Staff, Studies and Analysis, to Ofc of AF Hist, subj: Review of Historical Study, 18 Nov 75, cited hereafter as Botticelli ltr.

21. End-of-Tour Rprt (S), Col. Arthur D. Thomas, USAF, 460th TRW, 10 Oct 66, p 5, cited hereafter as Thomas Rprt.

22. Pierson, Electronic Warfare, pp 103-104.

Chapter III

1. Porter, Air Tactics, pp 11-14; Msg (TS), CINCPAC to JCS, subj: SA-2 Strike Planning, 2521152 Jul 65; Msg (S) CINCPAC to CINCPACAF, subj: Analysis of AAA Defenses Encountered in Rolling Thunder (Spring High), 2923442 Jul 65.

2. Vice Adm Malcolm W. Cagle, USN, "Task Force 77 in Action off Vietnam, " U.S. Naval Institute Proceedings, May 72, p 76, cited hereafter as Cagle, "Task Force 77; JCS Prong Tong Study, vol III, tab E to app 2 to annex L, pp 1-3.

3. Porter, Air Tactics, p 14.

4. JCS Prong Tong Study, vol II, tab A to app 3 to Annex B, pp 4-5.

5. Porter, Air Tactics, p 18; Cagle, "Task Force 77," p 76.

6. JCS Prong Tong Study, vol II, app 3 to annex B, pp 1-3; Corona Harvest, Electronic Warfare, pp 57-59.

7. Final Rprt (S), TA WC and APGC, Wild Weasel I (Eglin Phase), Bee 65, pp 3-5.

8. Ibid., pp 7-12, 84.

9. Ibid., pp 69-73; Atch (S) to memo, Maj John D. Shmoldas, USAF, for AF I CHO -- Attn Mr. Berger, 15 Jun 76, cited hereafter as atch to Shmoldas memo.

10. Final Rprt (S), APGC and TAWC, Wild Weasel I (Southeast Asia Phase), Mar 66, pp 1-3.

11. Ibid., pp74-77.

12. Ibid., pp 30, 75.

13. Ibid., pp 78-81.

14. Ibid., pp 112-113.

15. Ibid., p 37-39.

16. Corona Harvest, Electronic Warfare, p 61; Atch (S) to ltr, Maj Jesse P. Wiggins, USAF, to Hq USAF/CHO, subj: Review of Historical Manuscript Titled Tactics and Techniques of Electronic Countermeasures against North Vietnam, 1965- 1973, 1 Jun 76, cited hereafter as Wiggins ltr.

17. Final Rprt (S), APGC and TAWC, Wild Weasel IA, Jan 66, pp 1-7.

18. AWC Professional Study no 4828, Wild Weasel: The Evolution 0 Unique Weapons System, Apr 72, pp 55-56, cited hereafter as Wild Weasel Evolution.

19. Ibid., pp 58, 63.

20. Tactical Doctrine (S), 355th TFW, 31 Dec 67, p B:5.

21. Wild Weasel Evolution, pp 95-96.

22. Atch to Shmoldas memo; Hist(TS), Dir/Ops, Jul-Dec 69, pp 225-226.

23. Atch to Shmoldas memo; Wild Weasel Evolution, pp 34-36.

24. Excerpt (S) from 7AF Weekly Air Intelligence Summary, 23 Apr 66, supporting doc no 13 to Porter, Air Tactics.

25. Wild Weasel Evolution, pp 66-67.

Notes

26. Ibid., pp 67-68; Hist{S), 355th TFW, Jul-Dec 66, vol I, p 46; Hist (S), 354th TFS, Oct 66, supporting doc no 17 to Hist, 355th TFW, Jul-Dec 66; Atch (S) to Hist, 355th TFW Ops, Jan-Jun 67, supporting doc no 13 to Hist, 355th TFW, Jan-Jun 67; Hist (S), 388th TFW, Jul-Sep 68, vol I, pp 22-23; Atch (S) to ltr, Col Peter Tsouprake, USAF, to CHO, subj: Review of Historical Manuscript re: ECM vs. North Vietnam Defenses, 16 Aug 76, cited hereafter as Tsouprake ltr.

27. Wiggins ltr.

28. Ibid.

29. Porter, Air Tactics, p 33.

30. Wiggins ltr.

31. Project Corona Harvest Input, Electronic Warfare Effectiveness (SEA), 1965-1968 (S), Hq USAFSS, pp 51-52; Hist (S), 355th TFW, Jul-Sep 67, vol I, pp 50-51.

32. Wild Weasel Evolution, pp 74 -75; Atch (S) to Shmoldas memo; Tsouprake ltr.

33. Intvw (U), Hugh Ahmann with Lt Col Merlyn H. Dethlefsen, USAF, 354th TFS, 20 Dec 71, pp 64-66, cited hereafter as Dethlefsen intvw.

34. Ed Blair, "A Man Doing His Job," Airman, Apr 69, pp 54-55.

35. Dethlefsen intvw, p 76; Atch (S) to ltr, Col Robert S. Johnson, USAF, to Hq USAF I CHO, subj: Review of Historical Manuscript titled: Tactics and Techniques of Electronic Countermeasures against North Vietnam, 1965-1973, 7 Jun 76; Atch to Shmoldas memo.

36. Corona Harvest, Electronic Warfare, pp 67- 68; Rolling Thunder Handbook, pp 55-61.

37. Thomas Rprt, pp 5-6.

38. Wild Weasel Evolution, pp 81-82; Maj Victor B. Anthony, USAF, The Air Force in Southeast Asia: Tactics and Techniques of Night<5Pe ions, 1961-1970 (Ofc of AF Hist Mar 73), p 169, cited hereafter as Anthony, Night Operations; Tsouprake, ltr.

39. Project Corona Harvest Input, Support: Tactical Electronic Warfare, Apr 68-Dec 69 (S), Hq PACAF, pp 2'J-23, cited here- after as Corona Harvest, Tactical Electronic Warfare, 1968-1969; Hist (S), 357th TFS, Jul-Sep 68, pp 1-2, supporting doc to Hist, 355th TFW, Jul-Sep 68; Tsouprake ltr.

40. Project Corona Harvest Special Rprt (S), Ryan's Raiders, Jan 70, pp 1-2, cited hereafter as Ryan's Raiders.

41. Anthony, Night Operations, p 192.

42. Hist (S), 388th TFW, Apr-Dec 67, vol I, p 23; Jan-Mar 68, vol I, p 39; Jan-Jun 68, vol I, p 17, vol II, pp 22-23.

43. Ryan's Raiders, pp 2-4, 6-8.

44. Ibid., p 15.

45. Hist (S), 388th TFW, Jul-Sep 68, vol I, pp 25-26; Oct-Dec 68, vol I, pp 23-23A, 27; Atch (S) to ltr, 388th TFW to 7AF, subj: BASSI/II Concept of Ops, 4 Sep 68, supporting doc no 32 to Hist, 388th TFW, Jul-Sep 68.

46. Excerpt from 7AF Weekly Air Intelligence Summary, 23 Apr 66, supporting doc no 13 to Porter, Air Tactics; Corona Harvest, Electronic Warfare, p 65; Atch to Wiggins ltr.

47. Corona Harvest, Tactical Electronic Warfare, 1968-1969, pp 20-22; Pierson, Electronic Warfare, pp 109-113.

Chapter IV

1. Special Rprt (S), Oct 66-Aug 67, Col Robin Olds, CO 8th TFW.

2. End-of-Tour Rprt (S), 8 Jul 67, Brig Gen William S. Chairsell USAF, CO 388th TFW, p 4; Atch (S) to ltrs, Col Edgar A. Gill Jr., USAF, Chief, Strategic Forces/C3 Div, Dir Programs to Ofc of AF Hist, subj: Review of Historical Study, 10 Nov 75, and Col Robert L. Rodee, USAF, Chief, Tactical Div, Dir Ops, to Ofc of AF Hist, same subj, 11 Nov 75.

3. I-list (S), PACAF, Jan-Dec 67, p 66; Burch, Tactical Electronic Warfare, D 18.

4. Advance Evaluation Note (S), Anti-SAM Tactics (C), Air Development Squadron 5, 3 Dec 65, pp III:1-10; JCS Prong Tong Study, vol II, app 5 to annex B, pp 3-4.

5. Thomas Rprt, p 5.

6. Hist (S), APGC, Jan-Jun 66, vol I, pp 45-47.

7. Msg (S), APGC to AFSC, subj: Preliminary Reports and Conclusions, 2109302 Jan 66; Corona Harvest, Electronic Warfare, pp 71-72.

8. Vampyrus Final Rprt (TS), 7AF, 16 Oct 66.

9. Excerpt (S), Rprt of Meeting of SAB at Eglin AFB, Fla, 14-16 Nov 67, supporting doc no 4 to Burch, Tactical Electronic Warfare.

10. Intvw (S), Maj Lyn Officer, USAF, and Hugh Ahmann with Lt Col Robert E. Belli, USAF, F-105 pilot, 29 Jan 7 3, pp 94-95, cited hereafter as Belli intvw.

11. Atch 1 (S) to Staff Summary Sheet, 7AF Dir I Ops, subj: Proposed Weather Criteria, 30 Oct 72.
12. Hist (S), 355th TFW, Jan-Jun 67, vol I, pp 106-107.
13. Corona Harvest, Electronic Warfare, pp 83-84.
14. Special Rprt (S), Oct 66-Aug 67, Col Robin Olds, USAF, CO 8th TFW; Hist (S), APGC, Jul-Dec 67, vol I, p 53.
15. Corona Harvest, Electronic Warfare, pp 77-78
16. Ibid., pp 78-79.
17. Text of Briefing (TS), The Employment and Effectiveness of Missiles and Guided Weapons in SEAsia, 18 Jan 73, Dir/Ops, JCS, pp 32-34, cited hereafter as Missiles and Guided Weapons in SEAsia; Botticelli ltr.
18. Working Paper 68/4 (S), 7AF Dir/Tactical Analysis, Bombing Accuracy in a Combat Environment, 10 May 68; Rolling Thunder Handbook, p 56; Rolling Thunder-Linebacker Comparison, tab 16, app I.
19. Rolling Thunder Handbook, pp 26-28.
20. Hist (S), 388th TFW, Jan-Mar 68, vol I, p 22; Apr-Jun 68, vol I, pp 19-20.
21. Corona Harvest, Electronic Warfare, pp 80-82; Pierson, Electronic Warfare, p 91.
22. Hist Data Record Card (S), DCS, Ops, 388th TFW, Oct- Dec 68, exhibit 3 to Hist, 388th TFW, Oct- Dec 68; Project Corona Harvest Input, Support: Research and Development, 1961-1965 (S), Hq PACAF, p 78 cited hereafter as Corona Harvest R&D, 1961-1965.
23. Atch 4 to McLaurin ltr.
24. Ibid.
25. Corona Harvest, R&D, 1965-1968, pp 78-79.
26. Atch 4 to McLaurin ltr.
27. Ibid.
28. Corona Harvest, Tactical Electronic Warfare, 1968- 1969, pp 27-28; End-of-Tour Rprt (S), Lt Col James L. Fournier, USAF, 7AF EWO, pt III.
29. Hist (S), 355th TFW, Jan-Mar 69, vol I, p 69; Apr-Jun, vol I, pp 70-71.
30. Corona Harvest, R&D, 1965-1968, p 76; Belli intvw, p 95.
31. Corona Harvest, R&D, 1965-1968, pp 79-80; Clay Thesis, pp 65-66.

32. Lt Col Bernard Appel, USAF, "Bombing Accuracy in a Combat Environment," Air University Review, Jul-Aug 75, pp 40-41.

Chapter V

1. Hist (S), 355th TFW, Apr-Jun 69, vol I, p 69.

2. Hist (S), 41st TEWS, Oct-Dec 69, p 8, supporting doc to Hist, 355th TFW, Oct-Dec 69.

3. Corona Harvest, Electronic Warfare, pp 46-48; Corona Harvest, Tactical Electronic Warfare, 1968-1969, p 12.

4. Hist (S), 355th TFW, Oct-Dec 69, vol I, np.

5. Atch 1 to McLaurin ltr.

6. Corona Harvest, Tactical Electronic Warfare, 1968- 1969, pp 7-8; Boticelli ltr.

7. Hist (S), 355th TFW, Jul-Sep 69, vol I, p ix.

8. Intvw, Maj Lyn R. Officer, USAF, and Hugh Ahmann with Capt James L. Hendrickson, USAF, EB-66 EWO, 31 Jan 73, pp 7-8.

9. Hist (S), 355th TFW, Apr-Jun 69, vol I, pp v, 119.

10. Atch 2 to McLaurin ltr; Summary of Air Op , Southeast Asia (TS), Hq PACAF, Dir I Tactical Evaluation, Apr 68, p 5 :11, Sep 6 8, p 5 :16.

11. Corona Harvest, Tactical Electronic Warfare, 1968- 1969, p 24; Hist (S), 355th TFW, Jan-Mar 69, vol I, p 37.

12. Belli Intvw, pp 39-40.

13. Ibid., pp15-16.

14. Ibid., pp 21, 38-39.

15. Ibid., pp 18-19, 29-30.

16. Hist (S), 388th TFW, Oct- Dec 71, vol I, p 32.

17. Belli Intvw, pp 25-26.

18. William R. Karsteter, The Son Tay Raid Annex (TS) to Hist, Aerospace Rescue and Recovery Service, Jul 70-Jun 71; Rprt (TS), Son Tay Prisoner of war Rescue Op, Joint Contingency Task Group, 18 Dec 70; Atch 1 to McLaurin ltr.

19. Mel Porter, Proud Deep Alpha (TS), (PACAF: Project CHECO, 20 Jul 72), pp 1-2, 23, 28, cited hereafter as Porter, Proud Deep Alpha.

20. Ibid., pp 25-26; Atch 1 to McLaurin ltr.

21. Porter, Proud Deep Alpha, pp 27-29; atch l to McLaurin ltr.

22. Porter, Proud Deep Alpha, p 31; Atch l to McLaurin ltr.

23. Porter, Proud Deep Alpha, pp 26-27; Summary of Air Ops in Southeast Asia, Dec 71, pp 5-B-2, 3; Atch l to McLaurin ltr.

24. Hist (S), 388th TFW, Oct-Dec 71, vol I, pp 30-31.

25. Porter, Proud Deep Alpha, cited above, pp 50-54.

Chapter VI

1. Washington Post, 9 May 72; Melvin F. Porter, Linebacker: An Overview of the First 120 Days (TS), (Hq PACAF, Project CHECO, 27 Sep 73), pp 1-2, 11, 13, cited hereafter as Porter, Linebacker; SAC Linebacker II Chronology, p 6.

2. Hist (S), 8AF, Jul 71-Jun 72, pp 355-358; Target Commentary (S), app IA to Hist, 72d SW (Prov), Jun-Oct 72; Edward W. Knappman, ed, South Vietnam: U.S. - Communist Confrontation in Southeast Asia, 1972-1973 (New York: Facts on File, 1973C pp 182-183; SAC Linebacker II Chronology, pp 7-17; Atch 1 to McLaurin ltr.

3. Knappman, ed, South Vietnam, cited above, pp 216, 220-223, 227, 240.

4. Intvw (TS), Col John J. Rosenow, USAF, and Maj Richard B. Clement, USAF, with Lt Gen Robert J. Dixon, USAF, Vice Commander 7AF, 1-2 Dec 70, pp 19-20.

5. Msg (S), 7 I 13 AF to TAC Dir I Ops, subj: Electronic War- fare Summary, 3104252 Dec 72.

6. Msg (S),USAFTAWC to 388th TFW, subj: ALQ-71 Guard Limits, 1519152 Sep 72.

7. Hist (S), 8th TFW, Oct-Dec 72, vol I, pp 50-51.

8. Ibid., Jul-Sep 72, vol I, pp 37-38

9. Fighter Weapons Employment Guide (S), Radar Strike Div, Hq 474th TFW, 17 Aug 72; Concept for F-111A Employment in Southeast Asia (S), Hq PACAF, 27 Jun 72, p I:4.

10. Text of Briefing (S), A Summary of the Concept of F-111 Ops in Southeast Asia from 28 Sep to 17 Oct 72, 7 AF pp 29-30.

11. Msg (S), 366th TFW to TA WC, subj: Current Pod Settings and Employment, 181012 Sep 72; Msg (S), 7 I 13AF to TAC Dir I Ops, subj: Electronic Warfare Summary, 310425 2 Dec 72; Porter, Linebacker, p 71.

12. Msg (S), 7 I 13 AF to TAC Dir I Ops cited above; Employment Tactics, 1972 (S), 388th TFW.

13. Rolling Thunder-Linebacker Comparison, Tab 16, app I.

14. Msg (S), 8th TFW to 7AF Dir/Ops, subj: Linebacker LORAN Ops, 1400182 Oct 72.

15. Burch, Tactical Electronic Warfare, p 41; Missiles and Guided Weapons in SEAsia, pp 19a-20

16. Hist (S), 8th TFW, Jul-Sep 72, vol I, pp 46-47; Msg (S), 7113 to TAC Dir/Ops, subj: Electronic Warfare Summary, 3104252 Dec 72.

17. Hist (S), 8th TFW, Apr-Jun 72, vol I, pp 37-39.

18. Msg (S), 7AF to AF Spec Comm Center, subj: Chaff Tactics, 0212052 May 72, exhibit 58 to Hist, 307th SW, Apr-Jun 72; Atch (S) to Ltr, Comdr, Det 66, AF Spec Comm Center to SAC Hist Ofc. subj: Chaff Evaluation Briefing. 26 Apr 73. text for slide 10. cited hereafter as Chaff Evaluation Briefing.

19. Daily Event Analyses. Electronic •warfare Evaluation (S). Freedom Train (Comfy Boy 13-72). Aug 72. AF Spec Comm Center. vol II. pp 19-21.

20. Hist (S). 8th TFW. Jul-Sep 72. vol L pp 46-47; Oct- Dec 72. vol I. pp 51-52; Porter. Linebacker. pp 44-45; Msg (S). 7/13 AF to TAC Dir/Ops. subj: Electronic Warfare Summary. 310425Z Dec 72.

21. Missiles and Guided Weapons in SEAsia. pp 42-43; Review (S). 7/13AF Linebacker Tactics. Sep 72. pp 31-32; Tab B (S) to Minutes. 7AF Fighter Conference. 18-19 Jul 72.

22. Missiles and Guided Weapons in SEAsia. p 27; Msg (S). 7AF to Dir/Ops. Hq USAF. subj: Linebacker Conference Lima VII and Mike VII, 121231Z Oct 72.

23. Hist (S). 8th TFW. Jul-Sep 72. vol I. p 56.

24. Atch (S) to Staff Summary Sheet. 7AF Dir I Ops. subj: Analytical Notes on Strike/Support Ratio. 14 Nov 72.

25. Msg (S). 7/31AF to TAC Dir/Ops. subj: Electronic War- fare Summary. 310425Z Dec 72.

26. Atch 1 (S) to Staff Summary Sheet. 7AF Dir/ Ops. subj: Proposed Weather Criteria. 30 Oct 72; 8AFM 55-2 (S). Bomber Arc Light Crew Manual. 1 Nov 73. p 4:1; Hist (S). 8th TFW. Jul-Sep. vol I. pp 46-47.

27. Porter. Linebacker. p 68.

28. Atch 1 (S) to Staff Summary Sheet. 7AF Dir I Ops. subj: Proposed Weather Criteria. 30 Oct 72.

29. Atch 1 to McLaurin ltr.

30. Missiles and Guided Weapons in SEAsia. pp 27-28.

31. Msg (S). 7AF to AF Spec Comm Center, subj: Chaff Tactics, 021215 Z May 72. exhibit 58 to Hist, 307th SW. Apr-Jun 72.

32. Tab B (S) to Minutes, 7AF Fighter Tactics Conference, 18- 19 Jul 72.

33. Hist (S). 43d TEWS. Apr-Jun 72. p 1; Employment Tactics, 1972 (S), 388th TFW. pp 4-5; Msg (S). 7 I 13AF to TAC Dir/Ops. subj: Electronic Warfare Summary. 310425Z Dec 72.

34. Missiles and Guided Weapons in SEAsia. pp 17-18.

35. Hist (S). 42d TEWS. Apr-Jun 72, pp 19-20, 22; Hist (S), 388th TFW. Apr-Jun 72, vol I, p 55.

36. Hist (S), 388th TFW. Oct- Dec 72, vol I. p 37.

37. Ibid., pp 42-43.

38. Ibid., pp 42-44, 46, 54; Msg (S), 7AF to 8AF. subj: EB-66 Support Requirements, 160818Z Jul 72, exhibit 12 to Hist, 388th TFW. Jul- Sep 72.

39. Msgs' (S). AFSSO Udorn to AFSSG 7AF. subj: Linebacker . Critique (26 Dec72). 261355Z Dec 72; subj: Arc Lite Day 9 Critique. 290925Z Dec 72. supporting docs no 36 and 38 to Hist, 7I13AF, Jan-Dec 72.

40. Encl (S) to ltr, CINCPAC to CINCSAC et al., subj: CINCPAC Ops Security Survey of Arc Light Missions in Route Packages 2 and 3 (S), 27 Dec 72, exhibit 20 to Hist, 43d SW. Oct- Dec 72.

41. 7I13AF Linebacker Tactics Review (S). Sep 72, pp 25-26; Hist (S). 388th TFW, Apr-Jun 72, vol I, pp 52-54.

42. Missiles and Guided Weapons in SEAsia. p 27; Msg (S). 7 AF to TAC Dir I Ops. subj: Electronic Warfare Summary, 310425 Z Dec 72.

43. 7/13AF Linebacker Tactics Conference (S), 20-23 Sep 72, p 24; Hist (S), 388th TFW, Jul-Sep 72, vol I, pp 19-20; Employment Tactics. 1972 (S). 388th TFW, pp 3-4.

44. Employment Concepts (S dg C), 17th -wild -weasel Squadron, 26 Oct 72.

45. 7/13AF Linebacker Tactics Review (S), Sep 72, pp 25-27.

46. Hist (S), 388th TFW, Apr-Jun 72, vol I, p 65; Jul-Sep 72, vol I, pp 46-47; Oct-Dec 72, vol I, pp 32-33, 61; Hist (S), 17th Wild Weasel Squadron, Apr-Jun 72, n p.

47. Atch 1 (S) to Staff Summary Sheet, 7AF Dir/Ops. subj: Proposed Weather Criteria, 30 Oct 72; Employment Concepts (S dg C), Hq 17th Wild Weasel Squadron, 26 Oct 72; Msg (S), 7AF to TAC Dir/Ops, subj:

Electronic Warfare Summary, 310425Z Dec 72; Msg (S), 7AF to CTS-77 et al, subj: SAM Calls, 1712 12Z Oct 72.

48. Missiles and Guided Weapons in SEAsia, pp 24-25; Atch 1 to McLaurin ltr.

49. Atch 3 to McLaurin Ltr.

50. Msg (S), 7AF to AIG 8306, subj: SAM Firing Wrap-up (9 Apr 72), 120825Z Apr 72, exhibit 55 to Hist, 307th SW, Apr-Jun 72, pp 58-59.

51. SAC Linebacker II Chronology, p 15; Hist (S), 8AF, Jul 71-Jun 72, vol I, p xxxiv.

52. Hist (S), 307th SW, Oct-Dec 72, vol I, pp 38-40.

53. Msg (S), 7AF to TAC Dir/Ops, subj: Electronic Warfare Summary, 310425Z Dec 72.

54. SAC Linebacker II Chronology, pp 59-60.

55. Msg (S), AF Spec Comm Center to AIG 8556 et al, Comfy Coat 72-156 (Bravo), subj: Significant EW Events, 18 Dec 72, 222300 Z Dec 72.

56. SAC Linebacker II Chronology, pp 59, 70, 74, 79-80.

57. Ibid., pp 64-65.

58. Ibid., p 81; Capt D. D. McCrabbe, USAF, B-52 EWO, quoted in Hist (S), 307th SW, Oct-Dec 72, vol I, p 57; Chaff Evaluation Briefing, text for slide 11.

59. SAC Linebacker II Chronology, pp 99, 103, 109.

60. Ibid., pp 95-96.

61. Ibid,, p 313; Quoted in Hist (S), 307th SW, Oct- Dec 72, vol I, p 63.

62. Hist (S), 307th SW, Oct-Dec 72, vol I, p 70; Capt Jameison, B-52 radar navigator, quoted in Hist (S), 307th SW, Oct-Dec, vol I, p 61; Capt Kordenbrock, B-52 pilot, quoted in Hist (S), 307th SW, Oct-Dec 72, vol I, p 69; Maj Sweet, B-52 radar navigator, quoted in Hist (S), 307th SW, Oct-Dec 72, vol I, p 66.

63. Capt Kordenbrock quoted as above; TSgt 01 Quinn, B-52 tail gunner, quoted in Hist (S), 307th SW, Oct- Dec 72, vol I, p 61; SAC Linebacker II Chronology, p 313.

64. SAC Linebacker II Chronology, pp 152-154.

65. Msg (S), AF Spec Comm Center to AIG 8556 et al, Comfy Coat 72-156 (Charlie), subj: Significant EW Events, 18 Dec 72, 222345Z Dec 72.

66. Msg (S), AF Spec Comm Center to AIG 8556 e al. Comfy Coat 72-156 (Alpha), subj: Significant EW Events, 18 Dec 72, 222300Z Dec 72; Chaff Evaluation Briefing, text for slides 11 and 36A.

67. Chaff Evaluation Briefing, texts for slides 24-26; SAC Linebacker II Chronology, p 167.

68. Msg (S), AF Spec Comm Center to AIG 8556 et al Comfy Coat 72-159 (Bravo), subj: Significant EW Events, 20 Dec 72, 242348Z Dec 72.

69. SAC Linebacker II Chronology, pp 143-144; Hist (S), 43d SW, Jul-Dec 72, vol I p 77; End of Tour Report (S), Lt Gen Gerald W. Johnson, USAF CG 8AF, Sep 71-Sep 73, p 90.

70. SAC Linebacker II Chronology, pp 191, 202-204, 227.

71. Ibid., pp 172, 224-225.

72. Ibid., pp 222-223, 227-230.

73. Ibid., pp 228.

74. Ibid., pp 238-239, 241

75. Ibid., p 315.

76. Ibid., pp 314-315.

Chapter VII

1. Air Force Internal Working Paper (S), Out-Country Air Ops, Southeast Asia, l Jan 65-31 Mar 68, Final Rprt, Jul 73, Project Corona Harvest, p 59, cited hereafter as Corona Harvest Conclusions.

2. Ibid., pp 16, 70.

3. Pierson, Electronic Warfare, pp 43-48.

4. Ibid., pp 50-51.

5. Atch 1 to McLaurin ltr.

6. Corona Harvest Conclusions, p 68.

7. Pierson, Electronic Warfare, pp 127-129

8. Corona Harvest Conclusions, pp 68-69; Hist (S), 388th TFW, Apr-Jun 72, vol I, p 65; Jul-Sep 72, vol I, pp 46-47; Oct- Dec 72, vol L pp 32-33.

9. Special Rprt (S), Oct 66-Dec 67, Col Robin Olds, USAF, CO 8th TFW.

10. Corona Harvest Conclusions, p 60.

11. Ibid., p 63.

12. Chaff Evaluation Briefing, text to slides 34a, 36a; Atch 1 to McLaurin ltr.

13. Corona Harvest Conclusions, p 61; Pierson, Electronic Warfare, pp 113-115.

14. Corona Harvest Conclusions, pp 152-153, 167-170; Botticelli ltr.

15. Corona Harvest Conclusions, pp 26-27; Encl (S) to ltr, CINCPAC to CINCSAC et al, subj: CINCPAC Ops Security Survey of Arc Light Missions in Route Packages 2 and 3 (S), 27 Dec 72, exhibit 20 to Hist, 43d SW, Oct-Dec 72; Corona Harvest Electronic Warfare, p 109; Corona Harvest, Tactical Electronic Warfare, 1968-1969, pp 7-8, 12-13.

16. Corona Harvest Conclusions, pp 15-16.

APPENDIX: Airborne Control of Fighters

1. PACAF Tactics I Techniques Bulletin, no 2 (S), 5 May 65, supporting doc 2 to Porter, Air Tactics.

2. Corona Harvest Rprt on Big Eye I College Eye (S), Aug 69, ADC, pp x-xii, cited hereafter as ADC Big Eye/College Rprt; Corona Harvest Input on College Eye Task Force (S), Apr 68- Dec 69, ADC, pp 1-4, cited hereafter as ADC College Eve Input.

3. Msg (S), CINCPAC to C SAF, subj: EC-121 Aircraft, 010117Z Apr 65, supporting doc 5 to Grover C. Jarrett, History College Eye, Apr 65-Jun 69 (S), (ADC, Ofc of Command Hist, Oct 69), cited hereafter as Jarret, College Eye; Capt Richard M. Williams, USAF, A History of Big Eye I College Eye (S), (Hq 552d AEW&C Wg, 1969), p 10, cited hereafter as Williams, College Eye History.

4. Williams, College Eye History, pp 18-19.

5. Hq PACAF Tactics and Techniques Bulletin no 35 (S), 8 Feb 66, EC-121:C Ops in SEA; Capt Carl W. Reddel USAF, College Eye (TS), (Hq PACAF, Project CHECO, 1 Nov 68), pp 11-12, cited hereafter as Reddel, College Eye.

6. ADC Big Eye/ College Eye Rprt, p 2:3.

7. Ibid., pp 2:2-3, 2:5-6.

8. Hq PACAF, Tactics and Techniques Bulletin no 35 (S), 8 Feb 66, cited above.

9. ADC Big Eye/College Eye Rprt, pp 2:7-8, 2:15-17.

10. Ibid., pp 2:10-11; Maj Lowell J.K. Davis, USAF, "The EC-121 Radar Platform in Combat, (S)" Aerospace Commentary May 73, p 109, cited hereafter as Davis, 11 The EC-121 Radar Platform. "

11. ADC Big Eye/ College Eye Rprt, pp 2:11-12.
12. Williams, College Eye History, p 14.
13. Ibid., pp 13-15.
14. ADC Big Eye/College Eye Rprt, pp xii -xvi. 6:12-14.
15. Ibid., pp 1:10-11.
16. Williams, College Eye History, pp 32-33, 159; USAF Combat Victory Credits in Southeast Asia (Ofc/AF Hist, 1974), p 16, cited hereafter as Southeast Asia Victory Credits.
17. Transcripts (S), Intvw with Capt Jerry Kaffka, USAF College Eye Track Force, nd, pp 4-5.
18. Ibid., pp 5-6.
19. ADC Big Eye/College Eye Rprt, pp I:28-34.
20. Ibid., pp I:37-38.
21. Williams, College Eye History, pp 22-23; Hist Record (S), Big Eye Task Force, 31 Dec 65, pp 7-8.
22. Williams, College Eye History, pp 23-24.
23. Ibid., pp 35-39.
24. Ibid., pp 116.
25. Hist Record (S), Big Eye Task Force, 30 Jun 66, pp 11, 13; Jarrett, College Eye, pp 52-54.
26. Jarrett, College Eye, pp 56-57; ADC Big Eye/College Eye Rprt, pp 6:12-13; Working Paper 67/10 (S): College Eye Task Force, 1 Jul 67, Hq 7AF, Tactical Air Analysis Center, supporting doc 13 to Reddel, College Eye.
27. Hist Record (S), Big Eye Task Force, 31 Dec 66, pp 8-19.
28. Jarrett, College Eye, p 90; Hist Record (S), 552d AEWC Wg, 31 Dec 71, pp 3-4.
29. Ibid., pp 96-100; Msg (S), ADC to CINCNORAD, subj: QRC Installation. at Z-209, 041000Z May 66, supporting doc to Hist ADC, Jan-Jun 66.
30. Jarrett, College Eye, pp 10 1-102; Hist Record (S), College Eye Task Force, 30 Sep 67, pp 4-5; ADC College Eye Input, pp 84-85; Williams, College Eye History, p 119.
31. Msg(S), CETF to 552d AKW&C Wg, subj: 24 Oct 67 Bravo Mission, 251100Z Oct 67; 552d AEW&C Wg Sentinel, 30 Jul 71, exhibit 28 to Hist Record, 552d AEW&C Wg, 30 Sep 71.

32. Msg (S), College Eye Task Force to 552d AEW&C Wg, subj: Realignment of College Eye Orbits, 280811Z Nov 67; ADC Big Eye/College Eye Rprt, p 6:13.

33. Williams, College Eye History, pp 172-175.

34. Jarrett, College Eye, pp 109-113.

35. Ibid., pp 116-117.

36. Maj Anthony J. Skiscim, USAF, "Rivet Top, " (S), ADC Communications and Electronics Digest, Jul 68, pp 35-39.

37. Final Rprt (S), TAC Test and Evaluation, Rivet Top Air- craft System, Mar 68, TA WC, pp 24-25; Jarrett, College Eye, pp 131-132.

38. Rprt (S), Relative Effectiveness, College Eye, Rivet Top, Big Look, 10 Jan 68, 7AF Dir /Tactical Analysis, supporting doc 14 to Reddel, College Eye.

39. Msg(S), CETF to 552d AER&C Wg, subj: CETF Capability, 010400Z Jan 68.

40. Jarrett, College Eye, pp 141-142, Hist Record (S), 552d AEW&C Wg., 31 Dec 71.

41. Jarrett, College Eye pp 116-130.

42. Intvw (S), Capt Carl W. Reddel, USAF, with Capt Richard M. Williams, USAF, College Eye Task Force, 14 Jul 68.

43. Williams, College Eye History, pp 28-30.

44. Ibid., p 28; Jarrett, College Eye, pp 72-73.

45. Williams, College Eye History, pp 97-101, 112.

46. Ibid., p 120; Hist (S), ALC, Fiscal 72, p 321.

47. Williams, College Eye History, pp 131-136, 156.

48. Msg(S), Gen Momyer personal for Gen Ryan, subj: North Vietnamese GCI Advantage, 291140 Z Jan 68, supporting doc 43 to Jarrett, College Eye; Hist Record (S), College Eye Task Force, 31 Mar 68, pp 7-8.

49. Hist Record (S), College Eye Task Force, 31 Mar 68, p 5.

50. Ibid., pp 8-9; Hist (S), ADC, Jan-Jun 68, pp 241-242.

51. Hist Record (S), College Eye, 30 Sep 68, pp 2-3; Hist (S), ADC, Fiscal 69, p 351.

52. Hist Record (S), College Eye Task Force, 30 Jun 68, p 3; Hist (S), ADC, Jan-Jun 68, pp 350-353; Msg (S), 552d AEW&C Wg to ADC, subj:

College Eye Update, 060017 Z Jul 68, supporting doc 10 to Jarrett, College Eye.

53. End-of-Tour Rprt (S), Col Floyd McAllister, USAF, Task Force Comdr, 20 Aug 70, pp 5-6; Hist Record (S) College Task Force, 31 Dec 68, p 2.

54. Jarrett, College Eye, p 142.

55. Ibid., pp 171-184; ADC Big Eye/College Eye Rprt, p 3:4.

56. Hist Record (S), 552d AEW&C Wg, 30 Sep 70, pp 23, 31 Mar 71, pp 5-6; 31 Dec 71, pp 3-4; Hist Record (S). College Eye Task Force, 30 Sep 70, p 3; 31 Dec 70, p 3; 30 Sep 71, p 3.

57. Hist Record (S), 552d AEW&C Wg, 31 Dec 70, pp 9-14

58. Rprt (TS), Son Tay Prisoner of War Rescue Op, Joint Contingency Task Force, 18 Dec 70, pp 45, 67, 69, J:1-3.

59. Hist (S), ADC Fiscal 72, p 326; Hist Record (S), Det 1, 552d AEW&C Wg, 31 Mar 72, supporting doc 7 to Hist, 552d AEW&C Wg, Jan-Mar 72; 30 Sep 72, supporting doc 7 to Hist 552d AEW&C Wg, Jul-Sep 72, Southeast Asia Victory Credits, p 22.

60. Employment and Effectiveness of Missiles and Guided Weapons in SEAsia, p 12; Employment Tactics, 1972 (S), 388th TFW, p 5.

61. Employment and Effectiveness of Missiles and Guided Weapons in SEAsia, pp 12-13.

62. Hist Record (S), Det 1, 552d AEW&C Wg, 30 Sep 72, pp 6-7; Msg (TS), AFSSO Udorn to AFSSO 7AF, subj: Linebacker Tactical Conference, 271100Z Jul 72; Southeast Asia Victory Credits, p 23.

63. Employment and Effectiveness of Missiles and Guided Weapons in SEAsia, pp 41-42.

64. Msgs (S), 7AF to CTF-77 et al, subj: Linebacker MIG Warning Information, 311200Z Jul 72 and 062302Z Aug 72, subj: Teaball/Disco Control Procedures, 190100Z Aug 72.

65. Special Instructions no 2 (S), 27 Aug 72, 7 I 13AF.

66. Special Instructions no 5 (S), 17 Sep 72, 7 I 13AF; Msg (S), 7AF to CSAF, subj: Linebacker Command and Control 160115Z Sep 72.

67. Msg (S), 7AF to PACAF, subj: Status of Teaball Weapons Control Center, 170815Z Sep 72.

68. Atch (S) to Staff Summary Sheet, subj: Revised Ops Order for Teaball Weapons Control Center, 16 Sep 72, 7AF Dir I Ops.

69. Southeast Asia Victory Credits, p 25; Hist (S), 388th TFW, Oct-Dec 72, vol I, p 35; Hist Record (S), Det 1, 552d AEW&C Wg, 31 Dec 72, supporting doc 8 to Hist, 552d AER&C Wg, Oct-Dec 72.

70. Msgs (S), 7 I 13IAF SSO to 7AF SSO, subj: Linebacker Conference 24 Dec 72, 241715Z Dec 72; Maj Gen Hughes to Gen Vogt, subj: Linebacker II Arc Lite Day 11 Critique, 310310Z Dec 72, supporting docs 35 and 42 to Hist, 7 I 13AF, Jan- Dec 72; Hist Record (S), College Eye Task Force, 31 Dec 72, p 5.

71. Davis, "The EC-121 Radar Platform," pp 107-109.

72. End-of- Tour Rprt (S), Col George W. Rutter, USAF, CO 366th TFW. 15 Nov 72, p 11, supporting doc 9 to Hist, 366th TFW, Jul-Sep 72.

73. Msg (S), 7AF Dir/Ops to CINCPACAF Dir/Ops, subj: Teaball Radio Relay Backup Capability, 051136Z Jan 73.

74. Davis, "The EC-121 Radar Platform," pp 109-111.

www.ingramcontent.com/pod-product-compliance
Lightning Source LLC
Chambersburg PA
CBHW031300090426
42742CB00007B/535
* 9 7 8 1 9 3 9 3 3 5 1 8 0 *